Top-Gun Project Managers

8 Strategies for Reaching the Top of the PM Profession

Richard Morreale

First Edition

Oshawa, Ontario

Top-Gun Project Managers: 8 Strategies for Reaching the Top of the PM Profession
by Richard Morreale

Managing Editor:	Kevin Aguanno
Copy Editor:	Susan Andres
Typesetting:	Peggy LeTrent
Cover Design:	Troy O'Brien
eBook Conversion:	Charles Sin and Agustina Baid

Published by:
Multi-Media Publications Inc.
Box 58043, Rosslynn RPO
Oshawa, ON, Canada, L1J 8L6

http://www.mmpubs.com/

Paperback	ISBN-13: 978-1-55489-113-9
Adobe PDF ebook	ISBN-13: 978-1-55489-114-6

Published in Canada. Printed simultaneously in Canada, the United States of America, Australia, and the United Kingdom.

CIP data available from the publisher.

Table of Contents

Dedication

This book is dedicated to my wife Linda, our children, our grand children, our great grandchildren, our 4 chocolate Labradors and the enduring spirit of my constant companion for 13 1/2 years—my black Labrador, Jake, who passed away on March 22nd, 2011.

Preface

Why I Am Qualified to Write This Book

Well, you probably think that I'm going to tell you that I'm qualified because I studied project management in university and the university I attended certified me a project manager . Well, I didn't. Or, I studied the Project Management Body of Knowledge (PMBOK), took the test, and got my project management professional certification from the Project Management Institute. Well, I didn't. Or, I became an Association of Project Managers PRINCE2 (Projects in Controlled Environments) Certified Project Manager. Well, I didn't. I'm not suggesting that these certifications are not good things to go after. They are very good in teaching you the basics, the tools and techniques you should use to run a project. These tools and techniques you learn as part of the certification process are essential to use in managing a successful project. However, I don't believe that these certifications, these tools and techniques, turn you into a project manager.

You become a project manager by the interpersonal skills you develop over the years from dealing with people, by the attitudes and behaviors you develop as you gain experience and by using the basic tools and techniques. You become a successful project manager by paying attention to things you learn in every job you have in your career. You become a project manager by practicing project management. You become a project manager by running projects and screwing up. You learn more from your screw-ups than you do from your successes.

I believe, as you'll see in this book, that successful project management is heavily weighted on the side of soft skills rather than the skills you learn in the certification process. When you mesh the skills you are taught in the certification process with the soft skills you learn in dealing with people in lots of different situations, you become that truly great project manager—that project manager than can truly say that he or she is in the top 10% of all project managers.

I still haven't told you why I think I'm qualified. First, before I started working on the Apollo Program, putting men on the moon (and getting them back), I had something like eighteen to twenty various jobs, which gave me loads of experience in dealing with people. They helped me develop the soft skills and the attitudes and behaviors I have. I still have a long way to go, but at least I'm on the path. Let me give you an example of the jobs before Apollo. In no particular order, they included

- Working as a stock clerk in the Winn-Dixie Supermarket in Bay St. Louis, Mississippi.

Introduction

Most projects fail.... Based on the criteria of cost, schedule and expectations, most projects end up costing more than they were supposed to, or they take longer to deliver than they were supposed to or, if they are delivered, they don't meet the client's expectations. We should use one more criterion to determine whether the project is a success—a positive experience for the project team. We, as project managers, have a responsibility to establish an environment on the project that provides that positive experience for the project team. I want people working on my projects because they want to, not because they have to. I want people enjoying themselves on my projects rather than dreading coming to work. Also, I hate using the term *failure* to describe the results I get from an unsuccessful project. So, I'm going to start calling projects I've been calling project failures "unsuccessful projects." The only projects I'll call failures are unsuccessful projects from which we don't learn something to help us not make the same mistakes again. Then, I'll see those kinds of projects as failures and don't mind calling them that. My goal every time I conducted an audit was to learn the reasons for the poor results and help the teams not to do them again.

I recently wrote an article about what it takes to be a great project manager, and I got the following comment from a woman who used to work with me years ago. "The thing about reading your article is that I start to smile, and I remember that work can be fun, and hard, and exciting, and strenuous, and rewarding, and fun." A project manager should strive to make work on her project fun, and hard, and exciting, and strenuous, and rewarding, and fun—and make the people on the project smile. I like the saying "work hard and play hard," which is how a project should be run. Let's work hard when it's time to work, and let's play hard when it's time to play.

Are you ready? Then, let's go for it.

Chapter 1

The Top Reasons Projects Are Unsuccessful

The Top Reasons Projects Are Unsuccessful

Over the years, I have been asked to review and audit many projects—projects showing signs of trouble, projects in what could be considered "terminal trouble," and projects that had already given up the ghost. I was asked to do this for several reasons. The first was to see if the projects showing signs of trouble were really in trouble and, if they were, what we could do to keep them from going terminal. The second was to see if the projects considered in terminal trouble really were terminal and if there was something we could do to save them. And the third was to see what lessons the organization could take away from the projects that had already given up the ghost, so the mistakes made on that project weren't made again.

Projects are unsuccessful for many reasons. What I learned, however, over the years of managing projects, speaking with other project managers, and reviewing and auditing projects was that the same problems showed up repeatedly. The following are many challenges that cause projects to be unsuccessful:

- Lack of agreed requirements
- Lack of proper planning
- Poor or lack of change control
- Inadequate cost control
- No agreed development process
- Poor communication
- Lack of focus
- Lack of commitment
- Poor management

Lack of Agreed Requirements

The lack of agreed requirements is the number-one reason projects are unsuccessful. And it doesn't matter whether you are developing something that will take you a month or take you a year. It's still the same. The lack of agreed requirements is the number-one cause of unsuccessful projects. Now, no project manager should start a team working on a project without first knowing what the client wants, but some project managers do. They start the project without clearly understanding what the client wants. The project starts without an agreed requirement, or if there is one, it is usually an inadequate one. People start working to deliver what they **think** the customer wants, and soon enough, they discover that they really don't know what the customer wanted, and they've wasted lots of time, effort, and money. The project starts on the wrong foot and continues out of step. Morale could be affected. It's just a bad scene.

I've audited projects where I've asked the project manager if I could see the requirements document, and he didn't even know where a copy of it was. Some have even told me that they didn't have one, or they didn't need one because the team knew what they should deliver, or the project was small or the team was small, and everybody knew what everybody else was doing, anyway. I guarantee that's a sure setup for failure. In most cases, the size of the requirements document relates to the size of the project. If the project is small, then the requirements document will probably be small. If the project is big, then the document will probably be big.

I audited a large program for a financial services organization in London. This program was supposed to end the use of all paper in the transaction and dealing of equities in the City of London.

About forty brokerage houses were involved in this program, and after talking to several of them, I learned that everyone knew what the program was and what it was about, they knew what it would do, and they knew how it would do it. The only problem was everyone knew something different, and the worse part about it was his or her systems had to be prepared for the program when it went live. After I finished my audit, I was asked to take over as project manager. I agreed and promptly brought in two people—a highly experienced technical person and a great project office person—to help make the project successful. I could trust both, and they had the right attitude.

During my career, I've managed many IT projects, both big and small, from about $600M to a few hundred thousand, but I've never coded a line of code in my life. Wait a second; I lie. Yes, I did. I made the ZX-80—anybody remember the ZX-80? It was one of the first personal computers in London. I went into Dixon's where it was sold and wrote code on it so it repeated my name. Neat, huh?

The first thing I did when I took over this financial services program was to have the team define and agree what the requirements document would look like—its content and format. The team thought I was nuts to have them "waste" their time like this. After all, they had the requirements to analyze. They had been doing that for about two years and still didn't have a set of agreed requirements. So we prepared and agreed on the content and format standards of the requirements document. Using the standard as the basis, we agreed what activities had to be achieved to develop the requirements document. We then put together a plan comprising those activities. Working to the plan, we prepared a requirements document and reviewed it with the brokerage houses. When completed, it was approved by the company building the product and the forty or so brokerage houses. This was done in about three months.

Once we completed the requirements document, we looked at various ways to carry out the program, decided it was too expensive to build what they wanted to build, and we agreed that the best thing we could do was cancel the program and look at it again. That success could have been achieved years before if they had done it right. Instead, they wasted the years they spent trying to define the requirements in an unstructured way.

Once the first project was cancelled, the company building the product looked at it again and decided to establish a new program of work. I wasn't involved in the new program. A new internal project manager was assigned, and the new program started. Many of the same mistakes were made again, and eventually, the new program was put to death, kicking and screaming. The outcome? The new program ended up six years late with no solution in sight. It was 13,200% over budget and cost the company building the product and the financial institutions in the city about 800M GBP, or $1.2 Billion. As was the case with the first program, the lack of good requirements analysis and documentation of those requirements on the second project were the notable problems that escalated into and caused all sorts of other problems, eventually doing in the project.

Look, when you are not talking about information technology projects, it's easy to see why you need requirements documents. It would be foolish to have a carpenter come to build a house for you on some land you owned, and the requirements were that you wanted a "real nice" house that's "big enough" and has "enough" bathrooms. Oh, and by the way, you want it finished in three months, and you want to know how much it will cost. You see how foolish that sounds? Well, the same thing can be said about requirements for IT projects. Attempt to build a system without a set of requirements, and you know you will be in trouble.

I discuss requirements in detail in Chapter 3.

Lack of Proper Project Planning

Most projects I've reviewed that are in trouble do not have a proper schedule in place. I've found projects with high-level schedules that haven't been updated in months. I've found projects where there is a schedule, but it is on the shelf and the project manager is running the project by what only could be called "seat of the pants" project management. Without a proper schedule, the team members don't know what they are expected to do and by when. They don't know how what they are doing fits in the overall project. They don't know, with any certainty, how long the project will take, what resources are required, how much time is required of the client, and loads of other stuff they should know to have a successful project.

The information that should be included in a proper schedule includes:

- What must be delivered on the project—the individual products that make up the project delivery.
- The activities required to develop and to deliver the products.

- A dependency network detailing how the various activities fit together.
- The resources required to complete the activities.
- The planned start and end dates of each activity.

To me, a schedule that includes this information is an excellent project schedule that I can work to and monitor, track, and report on achievement. By the way, speaking of monitoring, tracking, and reporting, I've stopped using the terms *progress meetings* and *progress reports* or *status meetings* or *status reports*. I've started having achievement meetings and preparing achievement reports. So now, I have weekly achievement meetings and prepare weekly achievement reports. My meetings and reports to the project board are also called project board achievement meetings and project board achievement reports.

The words *achieve* and *achievement* affect those associated with the project differently from the words *progress* and *status*. In the future I'm going to call my project teams "achievement teams" and perhaps even call myself the "achievement manager." I admit that I haven't called my project team an achievement team or myself an achievement manager before, but if I get a chance to, I will.

When I consulted at one of the largest banks in London, I was asked to work with and coach a company director responsible for directing one of their major projects. I was talking to him about reporting when he told me that he had to distribute a progress report to the board the next day. I said that was great and asked why he didn't change the title of the report from project progress report to project achievement report. His first comment was something I didn't want to hear, and he shouldn't have been happy saying it. But, he said, "I don't think we really have achieved anything this month," which should have hinted how his project was going.

I was reviewing the IT projects for a large utility company in southwest England. I was speaking to a project manager, and I asked to see his project schedule. He told me that he knew he had one somewhere, and when he finally found a copy of a Gantt chart in the bottom drawer of his desk, I saw it hadn't been updated for about two months. If you are going to have a schedule that much out of date, you might as well not have a schedule at all.

I asked a programmer what was her role on the project and when what she was doing was needed. She told me she was just a programmer (telling me something about the attitude toward people in the company—**just** a programmer) and that the code she was producing had to be ready as soon as possible. That's no scheduled date. How can the people waiting for the code so they can do their job plan their work? OK, people, here's the schedule. We will start on the project as soon as possible, and by the way, we will finish the project as soon as possible. Great schedule, isn't it?

I use the project schedules we prepare to do more than just tell us how long the project will take. I also use them for monitoring and controlling, to gain commitment of the team, and to gain commitment of the bosses, the business, and the users. I also use them to help instill a teamwork culture in the project team, to help me get resources, to identify how the required date can be met, and to fight arbitrary end dates.

With a well-prepared project schedule at the right level—when senior management says that they want a project delivered on a specific date—we can show them what resources we need to meet that date. And if we are stuck with not enough resources, we can show them what date we can meet. If they continue to push us for an unrealistic early date, even when we show we can't meet it, then we must ask what they would like us to take out of the project requirements to make it feasible or what would they like us not to do in development. You can't put ten pounds of stuff in a five-pound bag. It will only hold five pounds, maybe six, and then catastrophe. It's the same with a project. You can't ask ten people to do the work of twenty people over the same period and expect it to get done.

Project planning is something we should be interested in throughout the entire project. We shouldn't just develop a schedule and put it on the shelf where it gets all dusty. One of the worst things a project manager can do is to develop a schedule on her own and then present it to her people as a "done deal." We need everyone's commitment to the schedule, and in my experience, commitment to the schedule comes from being part of the planning process.

I see the schedule as a document that should be updated and maintained on a timely basis. On a major program for the United Kingdom home office, we updated the achievement schedule with completion data weekly, and we had somewhere close to three thousand five- to ten-day activities on the schedule. We didn't look at every one of those activities weekly. We only looked at the activities that were supposed to start and those that were supposed to be completed during the week. Because we had the activities linked in a dependency network, if something were to go wrong with any activity, we could see the impact on succeeding activities.

Project planning is discussed in detail in Chapter 4.

Poor or Lack of Proper Change Control

Another major reason that projects are unsuccessful is because of poor or lack of proper change control. I have never asked a project manager, "Do you control changes?" and had him or her say, "Nah! We just let those suckers fly." They always say they control changes. But, in almost all cases, they don't control changes as rigorously as they need to. So, when I ask for the agreed baseline documentation, say, the requirements or design specification, he or she usually can't produce it. So I wonder what they are writing changes against—the ozone?

What seems to be the problem then? Why do most projects not have strict change control processes and procedures in place? What I've found is that most do have processes and procedures in place, but most project managers are not as insistent on the achievement team following them as they should be. In addition, what I've also found is that most change control procedures are bureaucratic and are followed creatively, if followed. I'm sure you know what I mean about being followed creatively. Anyway, what we need to do is be creative in writing the procedure and bureaucratic in following it. The definition I use here for *bureaucratic* is strict. Therefore, we must be creative in designing our change control process and strict in following it.

I've seen, as I'm sure some of you have, many situations where changes were made without impact assessment being conducted, and the change caused project slippage, increased costs, and technical problems. Proper change control is one way to stop these problems.

Let me give you an example of what happens when you don't have proper change control in place. When I was appointed the system development manager for a large insurance company in Liverpool, England, I found that people changed the live system without documenting or assessing the changes for cost, schedule, or technical impact before the changes were made. And why were they making them to the live system on the fly anyway? One problem we had because of that was many of our clients having their insurance premiums debited from their bank account two and three times a month, which was a mess. The IT people didn't do that after I took over.

People are generally nice, and they do want to provide great service to their clients. So, when a client asks the project manager or a member of the achievement team if "just this small change" can be made without having to go through that "bureaucratic change control process," the project manager or the team member, who wants to provide great service, makes the change. Before you know it, many changes have been made to the requirements without going through proper change control, and the project has lost its focus. Though the people doing the change think it is a small change, without a proper impact assessment, they don't know how the change affects the project cost, the schedule, or the technical side of things. Some changes, though they might not seem to affect the project negatively, cause problems.

Change control is discussed in detail in Chapter 8.

Inadequate Cost Control

Another major reason for unsuccessful projects is inadequate cost control. Most projects I reviewed did not track costs against achievement. They did not do cost/schedule reporting. It's great to be right on schedule, but it's not so great if you've had to spend twice the money you were supposed to spend to stay on schedule. You ask a project manager if he is on schedule, and he says he's right on it. Right where he's supposed to be. Great, then you ask him this, how much money has he spent to be on schedule? Most of them hem and haw and admit they are unsure, or when they are sure, most of them have spent more than they should have. Most successful managers track both cost and schedule.

Cost control is discussed in detail in Chapter 8.

No Agreed Development Process

Too often, project managers use the old seat-of-the-pants method to manage their projects. That is, the project is started, and it continues to be run without an agreed development process in place.

One of the biggest problems with this is that the project manager is the only one who knows what the overall plan is. At least, he thinks he knows what the overall plan is. But these projects usually do not have a good project plan in place, either. The project manager is making it up as he goes. And that causes big problems. Communication is usually a big problem on most projects, even the ones run successfully. When you add a project run without an agreed development process in place, you have big problems. For a successful project, the people on the project need to know beforehand how the project will be run and, in addition, they should not only know what the development process is, they should have a hand in helping define it.

You can probably see why a company not having a standard development process in place can cause many problems. For instance, say your company has ten project managers and each is doing his or her "thing." They don't have a standard way of planning, a standard documentation set for their projects; they don't have a standard way for project reporting, no standard for project meetings, for monitoring the project, for estimating, and on and on. Every time a new project starts, the team is not sure how to do the job, or worse yet, every member of the team knows how to do it—just differently. It also makes it very difficult for senior management to review project achievement if each project does project reporting differently. It's also much easier to estimate project activities if a standard process is in place.

A standard development process in place makes it easier for the project schedule to be developed as the products, activities, and deliverables are already identified. That's half the planning job right there. You then need to identify the resources, estimated effort, and the elapsed time for each activity, get everyone's commitment to the schedule, and you're on your way.

Project planning is covered in Chapter 4.

Poor Communication

Communication is usually a big challenge, even in well-run projects. And, for sure, poor communication is a major reason projects are unsuccessful.

I've never seen a communication survey done at any company where most surveyed say, "Hey! Communication is just great. We all know what's going on." That just never happens. What's usually said is that communication is poor and needs to get much better.

I've seen it so many times, and maybe you have where people sit around the table in a meeting talking about poor communication. We need to communicate better. We need to make sure our people know what is going on. Yeah, we need to do better. Let's do it, people. Then, we walk out of that meeting feeling good about ourselves because we've recognized the poor communication and agreed that we will do something about it. Yeah, right. It just never happens.

You must put a plan in place to attack poor communication. You can't just say that you will do something about it, or you will do better. When I was program director at a large bank in London, we hired a consultant to help us attack the communication problem. He had us put a plan in place. The plan included quarterly off-site meetings with all project managers, monthly briefings presented by each director with the people in each division, monthly internal newsletters, a projects database where each project's status was covered in some detail, and a visibility wall so everyone could see project achievement, if he or she was interested. We tried communication by just saying we had to do a better job at communicating, and it didn't work. But our effort with the communication plan did.

Poor communication causes team members not to know what the project schedule is. Poor communication causes the project to lose focus. Poor communication causes team members to make decisions on the project where they do not have the authority to make the decision. Poor communication causes the development team not to develop what the client wants. Poor communication causes gossip and assumptions on the project. The project needs a communication plan, and the team needs to stick to it so that communication is a success multiplier and not a failure.

Lack of Focus

Lack of focus is another thing that pushes projects off the rail. Often, we start with laser focus on what we should be doing, and then things change. We start to lose that focus. Clients ask us to do things a little differently from the way the requirements say it should be done.

We're nice people. We don't want to say no, so we do it. It's a little thing anyway. Then, other things get in the way. The enthusiasm we had in the beginning for the project starts to wane. We get enthusiastic about other things that come up. We have things that come up outside the project we have to do.

Before you know it, you and your team have lost the laser-like focus you had on the project at the beginning. And once you lose the focus, it is difficult to get it back again. You can, but it would have been better not to lose it in the first place. The project manager needs to do everything in his power to keep non-project demands away from his team and keep his team focused on what it is the project is delivering.

Lack of Commitment

Lack of commitment is a killer. Most people are involved, not committed. You know the old example to show the difference between commitment and involvement. Take your bacon-and-eggs breakfast you might have had this morning. For that breakfast to be a success, the pig was committed while the chicken was only involved. Commitment is jumping in the pool to see whether it's cold, whereas involvement is just sticking your toe in the water.

When you have something to do, make sure you know what *it* is, and then just go for it. You don't want to go for it and not know what you are going for.

So, who has to be committed? You, your peers, your project team, the business, the users, the project sponsor, and all stakeholders on your project. A stakeholder is anyone who has a part in the project's success and/or will be using what is being delivered..

I want to talk a bit about the project sponsor. I have found that many times, the project sponsor is appointed to be project sponsor because he is the least busy senior manager. That's no good for the project. On my projects, I want a project sponsor that will "die in the ditch" for the project.

I once managed the development of a point-of-sale system for the sales director at a large insurance company in England. His idea was to have a laptop that came apart, and you were to give one part to the customer, and the salesman would keep the other part. The customer could then see everything the salesman was doing on his screen, and he wanted this all portable. As it turned out, the salesman would have to carry about fifteen batteries in the trunk of his car to last him the entire day. And at night, he would have to charge them for the next day.

Anyway, I was presenting to the board, talking about problems and possible solutions, and I was put through the wringer. Did the sales director say anything? No. Though he and I had talked about how we would handle this with the board, he left me hanging, slowly swinging in the breeze. That was the last thing I did for him. Instead of being a project sponsor who would die in the ditch for the project, I wanted him to be a project sponsor who got pushed in the ditch.

Poor Management

Poor management is reflected in many other reasons for failure. However, I point it out here on its own because many projects are assigned to people who don't have the experience to run projects. They might be the best technical person, and for that reason, they are assigned the project to manage. They might have just come off a project management training course, and although they know project management tools and techniques, they don't have the experience or the training in the project management soft skills, and therefore, they lack the experience or knowledge to run the project. They are set up for failure. And because of this inexperience and knowledge, they are sometimes coerced into agreeing to project budgets, time-scales, and so on without having the detail they need to agree to the budget or schedule.

Project managers need to be trained in both the project management hard and soft skills, and they must be assigned smaller projects until they have the necessary experience with the hard and soft skills to move up to managing larger projects. The project manager must understand and be proficient in many other skills, and we cover them throughout this book.

Chapter 1: Points to Remember

Points to Remember

Projects are unsuccessful for many reasons, but some reasons show up repeatedly:

- Lack of agreed requirements
- Lack of proper project planning
- Poor or lack of change control
- Inadequate cost control
- No agreed development process
- Poor communication
- Lack of focus
- Lack of commitment
- Poor management

The project manager must do whatever necessary to ensure that her project does not have any possibility of failing from these typical reasons and any others for that matter.

Time to Do Some Work on the Reasons Projects Are Unsuccessful

1. Do you believe there are other reasons projects are unsuccessful? If you do, list them below.

____

____

____

____

____

____

____

____

____

____

2. Have you ever been involved in what could be called an unsuccessful project—one costing more than it should or took longer to complete than it should or if it was delivered, did not meet the client's expectations? Spend some time thinking about this and then complete the following.

Name of Project ______________________________

Situation ___________________________________

____

____

____

____

____

____

____

____

____

3. Which Failure Criteria Were Affected?

Cost? ____
____
____
____

Schedule? ____
____
____
____

Expectations? ____
____
____
____

4. Which of the nine reasons for failure (or one of yours) applied to the project failure? Or was the reason something other than one of the nine. List the reasons below and what you will do to keep this situation from happening in the future.

Reason for Failure	Corrective Action for Future Projects

CHAPTER **2**

The Success Equation

History Lesson

Let's look at a little history from the late seventies. Surveys in the late seventies showed that about 70% of all IT projects were unsuccessful when judged against the criteria of cost, schedule, and expectations. My added Criteria of a positive experience for the project team wasn't even included in the survey. However, my estimate on the percentage of unsuccessful projects, if positive experience was included, and remember, it's only a guess, would be about 95%.

So what did we do? We spent loads of money. We spent wheelbarrows full of money. Over the last twenty to thirty years, we spent millions, if not billions, of dollars to fix this. What did we spend it on?

We spent it on getting our IT staff certified as project management professionals by the Project Management Institute. According to the Project Management Institute, as of September 2010, more than 550,000 IT professionals have become project management professionals worldwide. We spent it on getting our IT professionals certified as PRINCE and PRINCE2 professionals. According to the Association of Project Management in the United Kingdom, as of September 2009, more than 200,000 IT professionals have been certified as PRINCE and PRINCE2 project managers. In addition, many colleges and universities offered degrees in project management and offered certification to project managers.

We spent it on tools, techniques, processes, procedures, and training in the use of the tools, techniques, processes, and procedures. We spent it on project management methods. We spent it on learning and carrying out Structured System Analysis and Design Method (SSADM) and managing the SSADM process. We spent it on PERT. Anyone know what PERT stands for? That's right. Program Evaluation and Review Technique. I Googled PERT when I was preparing this chapter just to make sure I remembered what PERT stood for and found that there were at least forty-four different definitions for the acronym.

When I worked at the Mississippi Test Facility (now called the Stennis Space Center) on the Saturn II program, part of the Apollo Program for NASA, there was, across from our office in the headquarters building, a huge office loaded with drafting tables and many people diligently working at those tables. Probably between twenty-five and fifty people were in that office. My boss told me that they were PERT specialists. They used to put out reams and reams of paper. I don't know how many forests were destroyed because of the Saturn S-II PERT Department. Then, one day, I went into work, and their office was empty. They were gone. No, not to another area of the site, but just gone. And, you know what, nothing bad happened. We still kept testing, we still kept static firing, and we still kept shipping S-IIs to the Cape.

Now, I'm not saying that PERT was useless. It is an excellent tool, as all tools are, in the right place and at the right time. But what I want to show is that PERT is one thing we spent loads of money on, and sometimes, it wasn't worth it.

We spent it on work breakdown structure, product breakdown structure, project planning, and control. How many people use automated planning tools? What about that spend? We spent money on training in cost/schedule monitoring and reporting, change control, configuration management processes and systems, project monitoring and reporting, risk and issue management, budget control, life cycle management, return on quality. Remember return on quality? Quality circles, project boards, dependency networks, software development processes, and I could go on and on and on.

So considering all the money spent over the last thirty years or so on project management training, tools, and techniques and considering all the IT professionals certified as project management professionals and PRINCE2 project managers by the Project Management Institute, the Association of Project Managers, and colleges and universities, the failure rate of projects in the 2000s had to be much lower than it was in the late seventies. So let's look at what surveys in the middle of the first decade of the twenty-first century found.

Surveys in 2007 showed that, using the criteria of cost, schedule, and expectations, about 70% of all projects fail. That's right, 70%. The same 70% that was unsuccessful before we spent all that money since the seventies. The same percentage as before we certified almost a million project managers. Hmmm! Doesn't that make you wonder?

Well, first, it shows that project success must be based on knowing more than the mechanics of project management tools, techniques, and processes. Knowing and being proficient in the tools, techniques, and processes are what I call the hard skills, and these hard skills are essential to project success. However, the project manager must have and practice many other skills to deliver success. So what things besides the hard skills are required? Well, let's look at the success equation:

20%HS + 80%SS + MOP + PMCoA = Success

HS stands for the hard skills the project manager needs. SS stands for the soft skills. MOP stands for being a master of paradox, while PMCoA stands for the project management cycle of achievement. All these together equal success. So let's look at each of these parts of the success equation.

20% Hard Skills

The 20% hard skills I talk about here are planning, organizing, monitoring, and controlling. These typical overall management tasks must be the project manager's tasks.

Planning

The project manager needs to understand what a good plan consists of and how to put that plan together. We talk about planning the achievement in Chapter 4.

Organizing

The project manager needs to know how to organize the team for the people required, how many are required, and the best team structure is for the project. We talk about organizing for success in Chapter 5.

Monitoring

The project manager must know what needs to be monitored on the project and how to go about monitoring and reporting on it. We discuss monitoring and controlling achievement in Chapter 8.

Controlling

The project manager must understand what needs to be controlled on the project, what techniques to use in controlling the project, and how to use the techniques. We cover controlling in detail in Chapter 8.

These hard skills are important, as they form the project's foundation. And, as with a skyscraper or your house, if the foundation is weak, the rest of the skyscraper or of your house cannot be built successfully or be expected to last any time.

The same goes for a project. If the foundation, that is, all the proper hard skills tools associated with excellent project management, isn't there, the rest of the project will be unsuccessful. However, as with an office building or your house, the foundation is just a small part of the overall building; the hard skills are just a small part of the overall project.

Over the last thirty years or so, the industry has focused on the hard skills—the things project managers need to know to do the planning, organizing, monitoring, and controlling. There were times when project management training included some training in things such as communication skills and team building, but most of the time, they were treated as add-ons to what made a project manager—those hard skills.

Well, history shows us what happens when a person gets trained in only the hard skills and then, because of that training, is certified a project manager. As we have seen with the statistics, there seems much, much more to project management than the hard skills taught in the certification processes.

In my career, I've worked for, and with, some great project managers. So I started looking at what made them great. They had the hard skills. They knew how to plan, organize, monitor, and control. On top of that foundation, they practiced specific soft skills. They believed in the old song, "It ain't what you do; it's the way that you do it." And, boy, did they do it well. So the second part of the success equation is 80% soft skills.

80% Soft Skills

I run seminars and speak at conferences all over the US, Europe, the Middle East, and Asia on project management and project management soft skills.

And one of the exercises I conduct at most of my seminars is for the audience to list the soft skills, attitudes, and behaviors they believe project managers need to have to be great project managers. Wherever I am in the world, I get almost the same basic list. Here are several soft skills from that basic list.

Enthusiasm

The Free Dictionary defines *enthusiasm* as "Great excitement for or interest in a subject or cause." Successful project managers are excited about the project they manage. They are interested in their team. They are excited about going to work. Not only are they enthusiastic, but also they need to establish an environment on their project that will help generate enthusiasm in the team members.

Ralph Waldo Emerson has been quoted as saying, "Nothing great was ever achieved without enthusiasm." Think about it. Isn't that true? Anything I can think of that has been achieved has been achieved because someone was enthusiastic.

I read somewhere that one Fortune 500 company cuts the budget on their projects up to three times. Why? Because they believe that in doing that, they end up with the true fanatics on the project, that is, the enthusiastic ones.

Now, I'm talking about knowledgeable enthusiasm. It would be ridiculous to go running east all happy and enthusiastic, jumping around like crazy, looking for a sunset. It won't happen. You will not find that sunset however enthusiastic you are. And, you don't need to be jumping around to be enthusiastic. Quiet enthusiasm is just as good.

Energy

Project managers need to approach their projects with vigor, power, and energy. Running a large complicated project is a difficult job. Sometimes, long hours are required. Most of the time, the challenge affects us mentally and emotionally, but there are times when it can be challenging physically also. A project manager needs the energy to go the long haul.

Remember, work hard and play hard. You must have enough energy to do both. Eat well; sleep enough; take care of yourself. Those will help you go a long way to having the energy you need to complete your projects.

(I could take some of that "eat well" advice.)

Commitment to Excellence

Commitment is an absolute decision to do something. Successful project managers commit to doing everything on their project with excellence in mind. They want and they work at ensuring that whatever they deliver in completing the project is excellent. They are committed to and not just involved in ensuring excellence.

A great project manager must not only be committed to excellence but must also be able to inspire his people to be committed to excellence. What I'm talking about here is that a project manager should be committed to producing the best product that can be produced while meeting the project's requirements and constraints. A project manager should be committed to being the best project manager. The project manager should also be committed to working with the team to help them be the best they can be.

Commitment to Success

Again, we are talking about deciding to do whatever it takes to make the project a success. In other words, the project manager, and by his example, his entire team have decided that nothing will stand in the way of their being successful. Nothing. They will do whatever it takes, legally and morally, of course, to make success happen—whatever it takes to complete their projects on time, within budget, meeting the clients expectations, and having a positive experience in doing all this.

I provided project management consultancy to the managing director of a large company in England, and we attended a presentation by a project manager of a project about six months late and way over budget. The project manager's purpose was to talk about how he and his team would bring the project back in line or as close to in line as possible.

The new plan he put up and presented was aggressive, but I didn't question him about it in front of the managing director. After the meeting, I went up to him and commented that the new plan was aggressive. I asked him whether he was committed to meeting the new dates. He looked me in the eyes and said he didn't think they could do it, but he would try, and anyway, he knew the managing director wanted to hear that. Where was the commitment there? They didn't make it, and the project was cancelled, because by that time, they had missed their window of opportunity in the marketplace with the product they were delivering, and their business case was no longer workable.

Good Communicators

Project managers need to be great communicators. That is, they have to be good at exchanging information among themselves and others. They need to do this in both oral and written communication. Successful project managers and good communicators need to know when to speak and when to listen. I'm unsure where I heard this, but we are built with one mouth and two ears, and we should speak and listen in the same proportion. When I say listen, I don't mean listen so you'll know when the noise stops, and you can say what's on your mind. I mean, as Stephen Covey says, seek first to understand, then to be understood. And to understand, you must listen.

Do you have trouble paying attention or listening? Let me give you a hint of what I must do sometimes when I catch my mind wandering while I should be listening. I focus on the speaker; I work to pay close attention; I make a noise now and then to show him that I'm listening; and if I'm doing all those things, and I still find my mind wandering, I repeat in my head what the person is saying. I make sure that I repeat it in my head, as it would be a little disconcerting and embarrassing if I were caught moving my lips. If you have trouble listening, try it. See if it helps you.

Passion

How about passion? Great project managers have passion for what they do—passion for managing their project, passion for working with the people they are working with, passion for being a project manager, passion for taking an idea and, in a matter of time, making it real, and passion about life, in general.

Approachability

Then, there is approachability. Project managers should be approachable. They shouldn't just sit in their office at their desk with their door closed, doing whatever they do. They should wander around, looking for ways to serve the people working on their project. I see project managers as servants there to serve their people.

You've heard about MBWA (Managing by Wandering Around), which project managers should do. Wander around. Ask how people are doing, if there are any problems, and how you can help.

I worked with an executive board at a large company in Manchester, England, and I suggested to them that they should get out of their offices and wander around some, get to know their staff, be a little sociable, let the staff see them. I went back about a month later, and a director stopped me in the hall. He told me that the MD had tasked each director to take turns wandering around, and he was scheduled to wander around that week. Now, that was not what I had in mind. It was a little stiff, but something was better than nothing.

Going the Extra Mile

The project manager must be committed to going the extra mile. Zig Ziglar, possibly the best motivator

I've ever met, says that there are no traffic jams on the extra mile. If you just show up on the extra mile, you are ahead of 85% of the people.

Here's a great example of going the extra mile. Over the last year or so, I've developed an allergy to some washing powders. If I come in contact with clothes or sheets washed in that soap powder, I break out in a terrible rash that itches like crazy. I can take medicine that stops the itching, and so forth, but that's not the point of the story.

Recently, I was booked to lead a Project Management Soft Skills Master Class at a large company in Cary, North Carolina. I was booked into an Embassy Suites hotel for the two nights of the class. I awoke in the middle of the first night with this terrible rash, itching like crazy. I took my medicine that slowed the symptoms some, although it made me tired for the first day of the class. In the morning, I told the desk clerk about what I had experienced, and he said they would see what they could do about it.

How did they handle it? They changed all the linen in my room to new linens that had never been used or washed and treated me to my dinner that night. Now, that was going the extra mile.

Here is another example. I just returned from my quarterly meeting with the National Speakers Association in Charlotte, NC. Whenever I am in Charlotte I stay at the Hilton Executive. I let them know the first time I stayed there three years ago that I needed linens that had never been used or washed. Now, whenever I book they always have my room ready with new linens that have never been washed. Now, that is going the extra mile.

Interpersonal Skills

Project managers need great interpersonal skills. They need to know how to get along with people. They need to know how to establish rapport, how to interpret what other people are saying, and how to interpret their body language. They need to know when to talk, when to listen, and when to shut up—all good things to know.

Positive Thinking

Great project managers should be positive thinkers. They should have a positive attitude and show it. In England, a common saying is "not too bad." The weather report says the weather's not too bad this morning. The traffic is not too bad on the M25 this morning. You walk in to the office in the morning with a big smile on your face, all raring to go, and say, "How is everybody?" And they say, not too bad. I tell them, "Hey! I fell off a ladder once, and I was doing not too bad."

Project managers should stand tall, head high, big smile. You're looking great. What, you don't feel that way? Then, do it anyway, and soon, you will feel that way. Thinking positively, by the way, doesn't mean that that's all you need to do to be successful. You can't just sit on your couch and think positively. You also must take positive action.

Go-for-It Attitude

How about a go-for-it attitude? Once we know what *it* is, let's go for it. If you try to do it before you know what *it* is, you might go 'frit'. Let's take massive action to achieve it. Again, let's make sure we do whatever it takes. Let's let nothing stand in our way.

No-Problem People

Project managers need to be no-problem people. A more positive way of saying it might be can-do people. Always look at why you **can** do it, not why you **can't** do it. If you've ever worked with no-problem people, you know what a wonderful thing it is. It feels great when you know that if you need something done, it will get done with no whining or complaining. And if it is so wonderful to work with a no-problem person, why not become one yourself?

Care for Their People

Project managers should care for their people. Respect them. Not in some false, pretend way, because people can see through that. But in a genuine way. Caring for their people doesn't mean to give them everything they want. It means to treat them fairly. It means to praise them when they do something well and criticizing them when they do something wrong. It means to be a servant leader. It means to think about how you can better serve them. What can you do to make their working life better? How can you help them do what they have been assigned to do?

Motivators

Successful project managers must understand how to motivate themselves as well as do what it takes to motivate their people. The way that project managers motivate their people is to create a positive environment on their project. If the right environment is established on the project, then the natural motivation the team members have inside them will manifest itself, and while operating in that environment, they will be constantly motivated.

Sense of Humor

The project manager should have a good sense of humor. The project manager's responsibility is to establish an environment where team members can laugh and enjoy themselves. The project manager must not take herself too seriously. If someone is the butt of a joke, she should make it her.

What the team delivers to the client is probably important and serious, and we should recognize that, but with a light atmosphere in place on the project, we have a much better chance of delivering the project successfully. So, project managers should lighten up, enjoy themselves, make sure the team members enjoy themselves and never, that's right, never take themselves too seriously. The project Manager and the team should have a great time!

A Little Light Matchmaking

I spoke about project management to a group of about forty project managers from around the world at an agricultural equipment manufacturing company. I was in this part of my presentation about all the soft skills project managers should have when this beautiful lady from Italy put up her hand. I asked her how could I help her, and she said that if I ever found a man with all these soft skills, could I let her know because she would marry him. I said I was sorry, but if I ever found that person I would marry either him or her, although my wife and children might have something to say about that.

OK, in finishing the soft skills, I am sold on these soft skills project managers need to be the best. Surely, you want to be the best. After all, being the best helps ensure success. For some of you more mature people reading this book, as an example, how many CDs do you think Tina Turner would have sold if one of her hit tunes were "Simply the Average," instead of "Simply the Best?" Not many, I'm sure.

Masters of Paradox (MoP)

One major part of the Project Management Success Equation is MoP, which stands for the Master of Paradox. Great project managers must be Masters of Paradox.

Big Ego—Small Ego

Great project managers need a **big ego.** That's right, a big ego. They need to believe that they can build anything. I've never built a bridge, but I know I could. I would hire the best civil engineer and put her in charge of the project's technical area. She would be my technical director. I would also hire the best bridge construction-planning person and put him in charge of the project support office. He would be my project support officer. The project manager, the technical director, and the project support officer—the 3 Amigos. And, although I would make the final decisions that had to be made on the project because I was the project manager, I would, if need be, ask what the technical director and the project support officer thought.

At the same time the project manager needs a big ego, he also needs a **small ego**. That is, he needs to give a large part of the credit for success to his team. He needs to make sure that the team delivering the project successfully gets the accolades they deserve.

Autocratic—Delegator

Project managers need to be **autocratic** in that they need to be take-charge people. They need to be able to act, that is, to do the task themselves, when needed, to get things done. At the same time, project managers need to be great **delegators.** They need to remember that they have a team working with them, and they are not the only people on the project who can do things right. They also need to understand that once they have delegated a task to someone on the team, they need to let the delegated person get on with it with minimal involvement from them.

Manager—Leader

Not only does the project manager need to be a great **manager**, she also needs to be a great **leader**. The **leader** decides the direction the team should be moving through the jungle and ensures that the team is motivated to travel through the jungle, while the manager sorts the work schedule and makes sure that the team knows the direction they are going, that they have a work schedule in place, and that they have the necessary tools to clear the jungle in the right direction. At the same time she is managing the project, she must also be doing some leading.

Complexity—Keep It Simple (KIS)

The project manager needs to deal with the **complexity** of some projects, at the same time, striving to **keep it simple**. So many times, project managers have made projects I audit so complicated unnecessarily. The project manager's responsibility is to make what is done on the project as clear and understandable as possible by eliminating as many complexities as possible. Some projects are complex, and the project manager must have the intelligence to manage the complex projects. However, at the same time, the project manager must continually strive to keep it simple. Remember KIS. It is usually KISS or Keep It Simple, Stupid, but I just wanted to be nice, which is why I left the "stupid" off.

Impatient—Patient

The project manager must be **impatient**, must manage with a sense of urgency, and must be impatient to get things done. At the same time, the project manager must understand that, usually, a patient project manager will establish the relationships on the project that will help make the project successful.

Ambiguity—Perfection

The project manager must be able to handle the ambiguity that some projects exhibit, at the same time, working to clear it up. The project manager should always be searching for perfection. Perfection will never be achieved; however, I'd rather shoot for perfection and not achieve it than think beforehand that I can't clear up the ambiguity and not shoot at all.

Big—Small

The project manager needs to see the big picture, in other words, see how her project relates to the client's or the company's other projects. At the same time, she must be able to concentrate on details—details she must understand, at times, to make good decisions.

The Project Management Cycle of Achievement (PMCoA)

The Project Management Cycle of Achievement is the last part of the Project Management Success Equation and comprises a process based on 8 Strategies that most successful project managers follow. They follow the Project Management Cycle of Achievement while making full use of their hard skills, soft skills, and Master of Paradox skills. The following is a graphical representation of the Project Management Cycle of Achievement.

Most successful project managers follow this Project Management Cycle of Achievement or something similar. The 8 Strategies starting at the top of the figure and following the arrows clockwise are the following:

- Know Your Outcome
- Plan the Achievement
- Organize for Success
- Gain and Maintain Commitment
- Take Massive Action

- Monitor and Control Achievement
- Stay Focused
- Have Fun

Each of these strategies will be covered in detail throughout this book with individual exercises that should help you decide how well you use these strategies and what you might be able to do to make yourself better at using them. Remember, as always, these exercises only work if you do them. By the way, these 8 Strategies can be applied by anyone to help him or her achieve almost anything. Let's take a top-level view of the 8 Strategies:

Strategy 1—Know Your Outcome

Know what you need to deliver before you start to work on delivering it. Know what you want to achieve before you start to achieve it.

Strategy 2—Plan the Achievement

Make sure you have an achievement plan in place for achieving the outcome—delivering what the client wants you to deliver. In addition, start to use success words with your team and with project stakeholders. Words such as achievement, success, completion, and so on make a significant difference in helping propagate the right attitude.

Strategy 3—Organize for Success

There's that success word again. You must organize you, your team, and all others into a workable organization based on the achievement plan developed in Strategy 2. This will go a long way to helping ensure successful completion of your outcome.

Strategy 4—Gain and Maintain Commitment

Rather than just jumping into the work necessary to achieve the outcome, I suggest that you ensure that all who need to be committed to the achievement plan and to the project itself are committed, and you have a strategy in place to keep them committed.

Strategy 5—Take Massive Action

Once the outcome is known, the achievement plan is prepared, the organization is in place, and the team and stakeholders are committed, you must take massive action to your achievement plan. Action on its own does not equal achievement. However, massive action to the achievement plan does.

Strategy 6—Monitor and Control Achievement

As you take massive action to the achievement plan, you must ensure that you monitor achievement and, if required, change the plan. In addition, it is highly important that you have the necessary controls in place to handle changes, risks, problems, and so on that could affect your achievement.

Strategy 7—Stay Focused

You must remain focused on delivery of the outcome according to the achievement plan. However, if changes are required, make them and then refocus on the new outcome or the new achievement plan.

Strategy 8—Have Fun

One of the most important things you, as a project manager, can do with your team is to ensure a positive atmosphere on the project. You need to ensure that working on your project is enjoyable or, better yet, fun. One of your goals needs to be to have people working on your project because they want to, not because they have to. I know that people having fun and enjoying themselves are much more productive than people who aren't happy with their job.

Chapter 2: Points to Remember

Points to Remember

- The successful project management equation is 20%HS + 80%SS + MoP + PMCoA = Success.
- A great project manager must be proficient in the hard skills of planning, organizing, monitoring, and controlling.
- A great project manager must be experienced in the soft skills of attitude and behaviors.
- A project manager must continue to work to become better at both the hard and the soft skills.
- A great project manager must be a Master of Paradox.
- A great project manager should practice the 8 Strategies of the Project Management Cycle of Achievement:
    - Know Your Outcome
    - Plan the Achievement
    - Organize for Success
    - Gain and Maintain Commitment
    - Take Massive Action
    - Monitor and Control Achievement
    - Stay Focused
    - Have Fun

"It's not what you do; it's the way that you do it."

Time to Do Some Work on the Project Management Success Equation.

1. Rate yourself (gut feeling) from 1 (poor) to 10 (best) in the project management hard skills.

Skill	Rating (1 to 10)	Comments
Planning		
Organizing		
Monitoring		
Controlling		

2. Select eight soft skills from the list of project management soft skills in this chapter or add soft skills you think you need to be a great project manager. Rate yourself (gut feeling) from 1 to 10 in the project management soft skills you select.

Skill	Rating (1 to 10)	Comments

3. Rate yourself (gut feeling) from 1 to 10 as how you score for each Master of Paradox entity.

Paradox	Rating (1 to 10)	Comments
Big Ego/Small Ego		
Autocrat/Delegator		
Leader/Manager		
Complexity/KIS		
Ambiguity/Perfection		
Impatient/Patient		
Oral/Written		
Big/Small		

4. Looking at the Project Management Cycle of Achievement and your project management, how do you rate yourself (gut feeling) in each of the 8 Strategies?

Strategy	Rating (1 to 10)	Comments
Know Your Outcome		
Plan the Achievement		
Organize for Success		
Gain and Maintain Commitment		
Take Massive Action		
Monitor and Control Achievement		
Stay Focused		
Have Fun		

5. From each of the four rating exercises, select the areas that would be good to work on (look for the lowest ratings) and create a personal development plan to describe the action you will take to raise your score in that area. Do this rating every six months to see how you have progressed.

Chapter 3

Strategy 1—Know Your Outcome

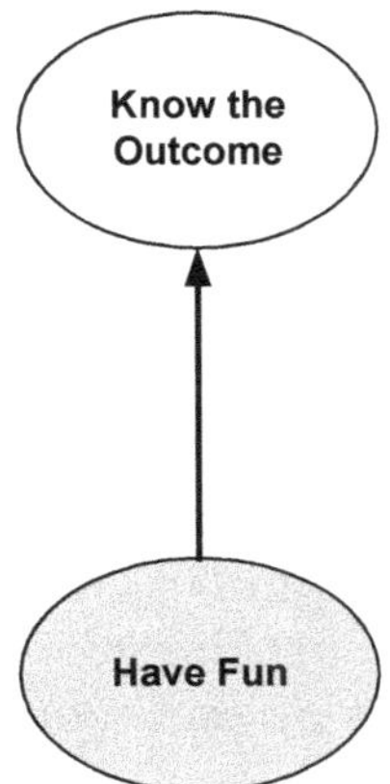

Know Your Outcome

The lack of an agreed, properly controlled requirement is the top reason projects fail. So, what do highly successful project managers do to help ensure success on their projects? They ensure that they have a properly documented, agreed, and properly controlled requirements document. This chapter focuses on the requirements' importance, the criteria for well-defined requirements, ways of defining the requirements, ways of handling unknowns, documenting the requirements, getting the requirements document approved, and controlling changes to it.

The Importance of Knowing Your Outcome

Having a proper requirements document in place on your project delivers many benefits to you, to your team, to the customer, and to the project's successful completion. Probably the major benefit of having the requirements document is that the project team will know what to deliver. There will be little wasted time associated with having to do things over because they started working on what they thought was wanted before they **knew** what was wanted. In addition, the requirements document provides a basis for evaluating, assessing, and agreeing changes to the overall project.

The requirements document also helps in getting commitment from both the project team and the person for whom the project is completed. It helps in pulling the project team together and in rallying the troops. It helps in the communication process because it provides a common basis for communicating what is wanted and, therefore, the basis of the project.

Lastly, the agreed requirements provide the basis for the project plan. I see no other way of putting a credible project plan together without having a definite idea about what must be delivered.

The Criteria for Well-Defined Requirements

A well-defined requirement must be a real document. It shouldn't just be a loose collection of e-mail, notes, letters, conversations, minutes of meetings, and so on. It should be a well-structured hard copy of a document signed by both the customer and the person responsible for delivery to the customer. Sometimes, it might be worthwhile to have the entire project team sign it as a sign not only of understanding, but also of commitment.

The requirement should be prepared by the customer and presented to the project team. However, the project team usually develops the requirements document with help from the customer. I'm OK with this if the customer supports the requirements document's preparation and signs it, signifying agreement that the document accurately defines what they want delivered, and they are committed to helping the project team deliver it.

Ways of Defining the Requirement

There are many ways to define the requirements. These include, but are not limited to, customer interviews by project team members, structured workshops facilitated by the project manager and attended by the customer and the project team, structured methods, prototyping, and visioning.

My favorite way to establish the requirements is to use structured workshops, especially when you incorporate into the workshop a structured technique such as a modified version of Structured

System Analysis and Design Method (SSADM) for documenting the requirement. I have used the Structured Workshop/Modified SSADM technique many times, and it has always proved to take less time to get to the requirements document than any other way.

Workshop/visioning is also an excellent way to define the requirements, especially when the project is for a training course or something else that doesn't necessarily require software development. Using the training course as an example, this is how it would work.

At the workshop, the facilitator leads the attendees in an exercise where they pretend they are in the future after the training has occurred. The attendees are then asked to identify what is different about the people who attended the training course. Not what they learned or were supposed to have learned in the training course, but what is different about them now? What can they do now that they couldn't do before? How do they act now compared with before the course? The attempt is to define, in detail, as much as possible the difference between now and before the training. The vision you produce is the requirement. It is documented and agreed as the final action at the workshop.

Whichever way you gather the requirements, you must put them in a requirements document that will be reviewed, agreed by, and signed by applicable stakeholders on the project. However, requirements are not the only thing that should be included in the requirements document. Acceptance criteria for each requirement should also be included.

Acceptance Criteria for Each Requirement

Each detailed requirement in the requirements section of the requirements document must have corresponding acceptance criteria in the acceptance criteria section of that document. There are many benefits relating to the project by including the acceptance criteria in the requirements document.

The first benefit is in identifying holes that might exist in the detailed requirements. We all know that errors found this early in the project's life cycle are much less expensive and much easier to correct than if they are found later in the life cycle, say, during testing or much later in implementation. The second benefit is that once the acceptance criteria are identified and included in the requirements document, the design process becomes much easier.

The third benefit is in testing. I like to set up an independent test function on my projects. We ensure that the requirements document is distributed to the independent test function so they can test what the product should do rather than what the product does. Hopefully, they are the same. The acceptance criteria help the independent test function in preparing their test plans, procedures, and scripts. We will talk about an independent test function in Chapter 5 of this book.

Ways of Handling Unknowns

I am well aware that not every detail of the requirement is known at the project's beginning. However, more of what are initially identified as unknowns are usually known or can be defined with little effort. In many projects I have audited, I have learned that the unknowns are usually forgotten until they come up later and cause many problems.

The way I handle things when I am told that some information is unknown is that I push back to make sure that they are unknowns. I then ensure that we define the unknown as much as possible. For instance, I want to know exactly what the unknown is. What is the effect on cost and schedule of not knowing the information required? Who needs to provide the information required? Where will it be found? When will we know it? I then document the unknowns in the project schedule and in the risks and issues, if applicable. I make sure that "unknown" information providers are aware of their responsibility, and I then monitor the unknowns closely.

Getting the Requirements Document Approved

You must get the requirements document approved by the applicable project stakeholders. The customer and project manager are the minimum people who should agree the document. In addition, I look at the other stakeholders on the project and decide who else should sign, based on their importance to project success.

Many people now get electronic approval of items. I never e-mail the final requirements document and ask for electronic approval. I know that electronic approval is a speedy way of doing things, but it is much better, if possible, to get approval face to face.

When I am ready for agreement and approval, I print a hard copy of the requirements document, and at a special meeting or meetings held to agree and approve it, I get the applicable people to sign it. Use e-mail to pass the document around for review and comment. But when it comes to getting it agreed and approved, print a hardcopy version and meet with those who must sign it to get their signature. I have even had a meeting of all the signatories where I presented the final document and then passed it around for signature. I've done this with business cases, requirements documents, project management plans, and so forth.

I like having meetings with stakeholders and signing hard copies of the final document because it highlights the document's importance and helps add to the attendees' commitment. I use these meetings with people to explain the importance of their involvement in the project and to ensure that they understand the requirement. I also want to make sure they understand what is asked of them and what they commit to by signing the requirements document. It is important that the people signing understand what their responsibilities will be, and a face-to-face meeting and a signature are good ways to make sure that they do.

One thing we don't do enough of on our projects is celebrate. So, once the document is signed, why not have a celebration? Pop open a bottle of champagne, mineral water, or something. When I was one of nine program directors of a major change program for a high street bank in London, we each had to get a business case approved once a year on the three-year program. Approval meant to get the signatures of many senior managers. All the program directors and various support staff met once a week on a Thursday morning from 8:30 to 10:30.

The day after I was the first successful program director to get my business case approved, I showed up at the Thursday meeting, announced that we had been successful in getting the business case signed, and my secretary hauled in a few bottles of champagne and orange juice. We celebrated the occasion throughout the meeting with several Bucks Fizzes. Judging how well the meeting went, we thought we should have done it more often.

Controlling Changes to the Requirement

Once the requirements document is approved, it should be base-lined and put under change control. That way, whenever a change to the requirement is necessary, it should be assessed for its technical impact on the project and its impact on schedule and cost. Once the impact is known, the person in charge of the requirement, normally the customer, can decide whether to approve the change. If he approves the change, he must accept the impacts identified during the assessments. If affected, the schedule, the budget, and the technical aspects of the project must be revised in line with the impacts. Spend the time getting the requirement right in the beginning and controlling it throughout the project, and you will save loads of time on your project.

Establish the Format and Content for Any Document to Be Prepared on the Project

This first strategy can affect other documentation required to be prepared on the project. Never start preparing documentation without defining, as much as you can, the documentation's outcome, in other words, the document's content and format standards. This fits right in line with knowing what you must deliver before you start delivering it. You should identify the document's structure in terms of the various chapters, sections, subsections, and so forth and what goes into each of these entities.

Agree this with your client or anyone else whose approval is required and then use this structure to produce your plan for the document's production. We'll cover planning in detail in Strategy 2. However, the planning process is to identify the activities required to complete each document entity, how long it will take to complete the activity, the order in which the entities must be completed (which depend on which), who must do each activity, and the scheduled start and end dates for each activity. I suggest that you do this to save time, to save wasted effort, and to ensure that you produce the best-quality documentation.

Chapter 3: Points to Remember

Points to Remember

- The lack of an agreed, approved, and properly controlled requirement is a major reason projects fail.
- There are many reasons the requirements document is necessary.
  - It forms the "contract" between the customer and the project team for what must be delivered.
  - It forms the basis for future development work.
  - It is the basis for the assessment of change on the project.
  - It forms the basis for the acceptance testing of what is delivered.
- The requirement should be prepared and approved by the customer and agreed by the developers. (Doesn't usually happen that way—usually the developers work with the customer to document, agree, and approve the requirements, when there are requirements documented and agreed, that is.)
- There are many ways of documenting the requirement:
  - Individual customer interviews
  - Structured requirements workshops
  - Prototyping
  - Visioning

- Include acceptance criteria for each requirement in the requirements document.
- Identify, document, and manage unknowns in the project schedule.
- Get the requirements document agreed, approved, and signed in person.
- Once agreed, put the requirements document under change control.
- Define the content and format of any of your project documents before you start producing them.

Time to Do Some Work on Strategy 1—Know Your Outcome

Answer the following questions.

1. Do you establish, document, agree, and approve the requirement for your project before you start to work on it? Yes? That's great. If you don't, why not, and will you do it?

____

____

____

____

____

____

____

____

____

____

____

____

2. What process do you, or will you, use to define the requirement with the client?

3. Do you now, or will you, include the acceptance criteria in the requirements document? If you do now, great! If you won't, why not, and where will the acceptance criteria be documented?

4. Do you now, or will you, document and manage unknowns in the project plan? If yes, great! If not, why not?

5. Do you now, or will you, get the requirements document approved, agreed, and signed in person? If yes, then great! If not, why not?

____
____
____
____
____
____
____
____

6. Do you now, or will you, put the agreed document under change control? If yes, then great! If you won't, why not?

____
____
____
____
____
____
____
____

CHAPTER 4

Strategy 2—Plan the Achievement

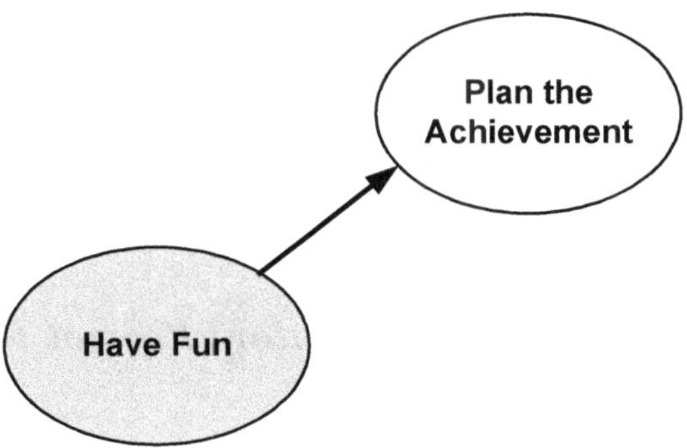

Introduction

The second strategy of successful project managers is plan the achievement. There are many reasons to plan the achievement. We know that without a project schedule we are never sure of what we should be doing, when we should be doing it, when we should have it finished, who depends on our finishing it, when the entire project will be completed, and how we will know whether we are on time for delivery. Those reasons are enough for project managers to pay particular attention to planning the achievement on their projects. (I use *project plan* and *project schedule* interchangeably.) Project planning is one of the most essential steps to take in the management of a project. It is only second in importance to having an agreed requirements document.

The Importance of Planning

One of my mentors (he doesn't know he is a mentor of mine but I have read so many of his books and listened to so many of his audio programs that I feel as if I know him), Zig Ziglar, says, "Failing to plan is planning to fail." Without a plan in place, there can be no control, with resulting chaos. We won't know where we are going or how or when we will get there. Success will just not happen.

I have had to answer many comments about planning over the years:.

> ***"I don't have time to plan. I've got to get on with the real work."***
>
> I can't understand what they are saying here. My question to them is what ***real work***? Without a documented plan, how do you know what real and productive work must be done.
>
> ***"What we plan now is bound to change."*** It will. This is even more reason we need a plan in place. We need an agreed plan in place so we can control the changes that will come. We need to consider the impact of proposed changes on the schedule, and we can only do that with an approved plan in place.

"How long will it take? How long is a piece of string?"

If they are going to say it, people usually say this when you ask them how long a particular activity will take. You can always estimate the time it will take to do any activity. Break it down into small pieces, estimate the time it will take for each of the small pieces, add the time together, and you have your estimated time to do the activity. Are you having trouble providing an estimate? The answer is to time-box the activity.

"Some things just can't be planned."

I have never found a project for which my team and I couldn't develop a plan. Later in this chapter, I'll Cover the planning process I always use.

When Does Planning Start?

Planning starts almost before anything else. Just the degree and the plan detail change as you move through the project's life cycle. For instance, as I've said many times before, it is impossible to put together a detailed plan for a project's delivery before you know what will be delivered. However, you can put together a detailed plan for the work required to identify and document what must be delivered, in other words, the activities required to prepare the requirements document.

When Does Planning Stop?

Planning stops after everything else has finished. Planning, with monitoring and updating of the plan, is a constant activity throughout the project's life.

Who Should Be Involved in the Planning?

If possible, most people involved in making the plan work should be involved in the planning in varying degrees. For instance, in the beginning of the project, the project manager should work with his team members to identify the development process that will be the plan's basis. The plan should then be prepared to reflect the agreed development process. Those stakeholders who have not taken part in preparing the plan should, at least, be made aware of the plan. Those team members joining the team after the plan has been prepared should have the opportunity to review and agree their assigned parts of the plan.

Some time ago, a large insurance company contracted me to lead a project that should have taken about three years to complete, but the project team had been working on it for about five years, and it still wasn't finished. The completion estimate then was for another two years, but no one trusted the estimate. As I found out later, the two-year estimate was strictly a "finger in the air" guess. The project was to replace all their existing IT systems, and it was about two years late when I took over. One of the first things I looked at was the existing plan, or I should say, the lack of an existing plan.

The project had degenerated into six teams working on several IT projects with each team managed by the old "seat of the pants" method with plans on the back of a cigarette pack. I thought that one of the first things I should do was to put a good structured plan in place so everyone knew what they should be doing, when they should be doing it, and who depended on their doing it. I also knew that I couldn't just dictate a process and a plan to them. First, I didn't know enough about the projects to do that, and if I knew enough and dictated a plan, they would not be committed to it. So I had to get them involved in defining the process and reflecting it in an agreed plan.

The six project managers, my project support manager, and I worked closely together over about two to three weeks and put a very good plan in place. During those two to three weeks, each project manager discussed the development process, the planning estimates, and so on with their individual teams and got their commitment. As the program manager, I challenged estimates and led the process and, at the end of the planning period, we had an excellent plan in place. The new plan covered a thirteen-month period that we stuck to for the rest of the project. The teams responded well to the new process and program of work, and we delivered on time.

What Comprises a Good, Detailed Plan?

Let's look at what comprises a good plan by looking at the criteria I use to decide if a plan is a good one:

Clearly Defined Objectives

A plan needs clearly defined objectives. That is, there needs to be a clear definition of what objectives are to be delivered.

Prepared with Monitoring in Mind

In other words, it must be deliverable-based and should identify the logical systematic process that must be achieved to produce and deliver the individual deliverables. A logical systematic process is very easy to monitor, and this monitoring helps the project manager in determining achievement to date.

Level of Detail

I plan the systematic process down to five to ten days' activities. I do this for many reasons. First, it is easier to estimate the effort for the project and the elapsed time for an activity if the activities are broken down to the five- to ten-day level. In addition, it is easier to monitor achievement of five- to ten-day activities regularly than it is to monitor activities scheduled to take longer.

All projects should be monitored weekly. When you monitor weekly, if a five-to ten-day activity goes wrong, you can do something to bring it back on track because you know about it early. If it were a thirty-day activity, and you learned it was going to slip, you couldn't be sure that you would know about it early enough to act to bring it back on track.

Some people think that planning at the five–ten level takes the plan down to too low a level. If that bothers them, what I propose now will probably drive them crazy. Some of my activities I monitor are not even one day in elapsed time—activities such as hand-overs, acceptance, and so forth. Let me explain the criteria I use for planning.

Criteria for Identifying Activities

I use several standard criteria to determine when I should establish a new activity. If the elapsed time for completion of the activity is estimated to be more than five to ten days, I find some way of breaking the activity down to a lower level of detail so the elapsed time is five to ten days or less. For instance, the activity is to produce a specification, and it has been estimated to take about thirty days elapsed time to produce it. To get down to at least a five-to ten-day level, I will break the specification down into five- to ten-day deliverable chunks, such as chapters, which will probably do the trick. If it doesn't, I then work to break it down even further.

I also establish a new activity when the responsibility for the work to be achieved changes. For instance, if we continue the production of the specification as an example, we find that the activity "Produce the Specification" is made of five activities: (1) produce the specification, (2) hand over to a group to review it, (3) review and comment on it, (4) incorporate applicable comments, and (5) issue it. The responsibility for each of the five activities changes. So, for planning purposes and to help me track where the specification is in the entire process besides ensuring clarity about who is responsible for what at various times in the process, I show the activity "Produce the Specification" as five activities in the plan.

Deliverable Product-Based Plans

Your project plans must be based on deliverable products. It is easier to manage the production of a deliverable product that you can see and feel than to manage many functional activities. Take, for example, analysis of the project's requirements.

During the Analysis Phase of the project, I could identify, in my plan, various activities that need to be completed to prepare things I might eventually need to be pulled together into a requirements document I will prepare later. I could also then have an activity in the plan, at the end of the stage, to "Produce a Requirements Document."

There are a number of problems associated with this way of planning and completing the work. Many times, once all analysis is completed, the document isn't produced. People say they will produce it later, but they never get around to it, and without it, they can't put the necessary scope-creep controls in place.

If the work is also done independent of the document production, how can you be sure that you aren't doing work that doesn't need to be done, and when you start putting your document together, you realize you didn't need to do some things you did? You can't be sure.

When you make your plans based on deliverable products, you know beforehand what the delivered product will contain. So you know that you won't be doing work that is not required. In addition, because the document is prepared as the work is achieved, there is no need to have an activity at the end to 'produce the document' as it is already produced.

Make all your plans deliverable product-based, and you will have a much better chance of success.

All Activities Must Help Prepare a Deliverable Product

Considering that the project is composed of deliverable products, all activities in the project plan, therefore, must help produce a deliverable product. If they don't, then, why are we wasting our time doing them? So, if you end up with activities called dangling activities, meaning activities not linked to an existing deliverable product, you should ask yourself several questions.

Why is this activity being done? What will be this activity's outcome? Will we get information we require from completing this activity? What deliverable product requires this information? Have we missed a product in our planning? With the answers to these and other questions you could ask yourself, you should then link this dangling activity to an existing product, add a new product, or delete the activity. It should not remain dangling.

Estimated Effort Compared with Elapsed Time

Here is another thing you need to remember about estimating effort. Effort does not equal elapsed time. You must look at how many productive days a week you can expect from someone working on your project. You need to consider ad hoc meetings, review of changes, requirements from other parts of the organization, sick leave, holidays, achievement meetings, and other things that could be considered overhead. On the British Income Tax System Project, COP, we planned for four productive days out of five days a week. So, if the effort on an activity were ten person-days, the elapsed time would be about two weeks and two days.

I worked to help plan another government project in the United Kingdom, and we discovered that project was achieving one productive day a week. I knew you weren't going to believe it. It was difficult for me to believe it also, but the figures didn't lie.

Dependencies, Resources, and Estimates

A good detailed plan must include not just the deliverables to be produced and delivered, not just the step-by-step activities that will need to be taken to achieve delivery, but also the dependencies among the activities, the resources required to achieve the activities, and the detailed estimated effort and elapsed time for completion.

The 7-Step Project Planning Process

The following is the 7-Step Project Planning Process that successful project managers follow to plan their project. The good project managers follow them without even thinking about them as separate steps. The 7-step Project Planning Process is shown below:

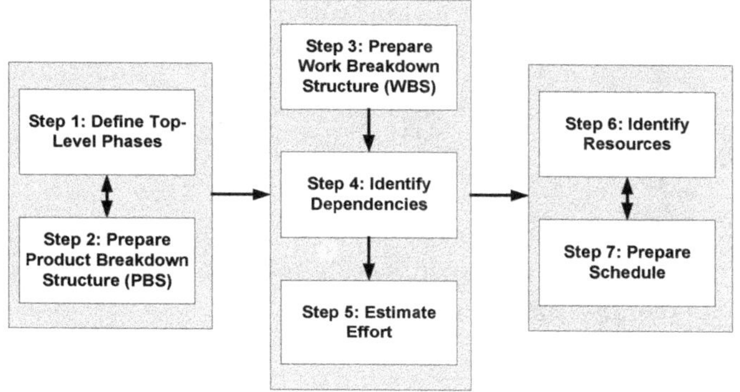

Step 1—Define Top Level Project Phases

One thing to remember about all things we talk about in the 8 Strategies is that, usually, terminology is unimportant. If the terminology is important, I will tell you it is important, but generally, what you call things is unimportant. What is important is that you are consistent in that once you name something, you should continue to refer to it as that. The reason I refer to that here is that I am now introducing the term *phase* to the planning process. A phase is a major part of the project life cycle, and the work comprising the phase produces at least one, if not more, major end products. Some people call a phase a stage. It doesn't matter to me what you call it. You could call it an axlelotle if you wish. Just be consistent, and always call it an axlelotle. For this book's purpose, I call it a phase.

The benefits of breaking the project down into major phases include:

- It provides an opportunity for a major review of the project.

- It provides an opportunity for a major review of the project's business case.
- It provides an opportunity for a major review and approval of the work done during the existing phase.
- It provides an opportunity to review and approve the proposed work to be done in the following phase.

Almost every project can be broken down into major phases of work. I haven't found one yet that can't. Let's look at some examples.

A project to build a house, for instance, could be broken down into four phases and major end-of-phase deliverables as follows:

Phase	End-of-Phase Product
Specification	Specification Document
Design	House Plans/Drawings
Build	Completed House
Acceptance	Owner-Inspected/Accepted House

Each phase meets the requirements for a phase in that it includes major effort, and each phase's output is a major deliverable.

A project to build an IT system, for instance, could also be broken down into nine phases and major end-of-phase deliverables as follows:

Phase	End of Phase Product
Project Initiation	Approved Project Initiation Report
Feasibility Study	Approved System
Options	Selected Options
Requirements Analysis	Requirements Document
Top-Level Design	Top-Level Design Specification
Development	Developed System (Code)
Testing	Tested System (Code)
Acceptance	Client-Accepted System
Implementation	Implemented System

These are just two examples, but you probably get the picture. This step is important because it forms the project plan's foundation. It forces you, from the beginning, to think logically about your project plan and, eventually, everything that must be done to deliver the project successfully. When done properly, it helps ensure that nothing is missed.

Step 2—Prepare Product Breakdown Structure

I suggest that one last activity you perform in any phase is to put the plan together at least for the next two phases. As part of the 7-Step Project Planning Process, step 2 says that you must identify the products to be developed and delivered during the completion and at the end of each phase. Therefore, if you are going to plan the next two phases, and I suggest that all your plans be product-based, you will need to identify the products for the next two phases as part of the entire process. Each phase will have at least one product, while some phases will have more than one.

I use a very loose definition of the term *products*. For instance, products can be management, technical, or quality items. Let me give you several examples of each.

Management products could be project management plans, weekly achievement reports, monthly achievement meetings, risk and issue reports, and so on. Technical products could be specifications, drawings, seating plans, design reviews, technical design meetings, and so on. Quality products could be quality management plans, quality reviews, quality review meetings, quality reports, and so on.

Once you have identified the products in the product breakdown structure, it is important to identify the content and format standards for each product. If you know before you start what you must deliver for what the product will look like, you will have a much better chance of defining and organizing the work to deliver it.

Step 3—Prepare Work Breakdown Structure

Once you've identified the products and the content and format standards for each product, it is time to identify the activities required to develop and deliver the products. Until you know what activities must be accomplished to develop and deliver the products, you can't identify the process you want the team to follow producing the product; you can't estimate the time required to complete the product, and you can't produce a detailed schedule for completion, which is why it is so important in this step to identify all activities for producing each product. In project management terms, this is called a work breakdown structure.

I suggest that you strive to establish, as much as possible, a generic series of activities to produce the products. Usually, I've used the following generic series of activities for every product at the product breakdown's lowest level.

- Prepare Draft
- Hand Over for Review
- Review and Comment on Draft
- Rework Draft with Appropriate Comments
- Agree Final
- Baseline Final

Let's say, for example, you are working with a requirements document broken down as follows:

- Introduction Chapter
- Existing Situation Chapter
- Functional Requirements Chapter
    - Requirement Number 1
    - Requirement Number 2
    - Requirement Number 3
- Data Requirements Chapter
- Acceptance Criteria Chapter
    - Requirement 1 Acceptance Criteria

- Requirement 2 Acceptance Criteria
- Requirement 3 Acceptance Criteria

Using this example, the generic activities used in the planning of the production of Functional Requirement Number 1 Chapter of the Requirements Document would be as follows:

- Prepare draft of Functional Requirement Number 1.
- Hand over draft of Functional Requirement Number 1 for review.
- Review and comment on draft Functional Requirement Number 1.
- Rework draft Functional Requirement Number 1 with appropriate comments.
- Agree final Functional Requirements Number 1.
- Baseline final Functional Requirements Number 1.

Using a set of generic activities helps make planning easier, helps make it easier for team members to remember the plan and to work with it and helps make it easier to track and report on achievement.

An example of using this set of activities in a different part of the requirements document is as follows:

- Prepare draft of Introduction.

- Hand over draft of Introduction for review.
- Review and comment on draft Introduction.
- Rework draft Introduction with appropriate comments.
- Agree final Introduction.
- Baseline final Introduction.

No matter what part of the requirements document you are planning, you can use the standard activities. Establishing a generic set of activities that can be used for most of your products and, therefore, for most of the plan is easier than it might sound. Try it.

Step 4—Identify Dependencies

Once you have the activities identified for each product in the work breakdown structure, the next step is to organize the activities into the order in which they must be accomplished. In other words, which activities must be completed before which other activities? Which sections of the products must be completed before other sections of the products? Which products must be completed before other products?

Organizing the activities, sections, and products into dependency order, in other words, producing a dependency network, will show you the process the project team needs to follow to complete the plan. This dependency network will show the entire team how the project will be completed, who depends on whom, what depends on what, and how it all fits together. Any team member can see how any slippage, or lack of delivery on his or her part, will affect other team members and the project.

Dependency networks are easy to do, as there are many software-planning packages to help draw the network. The important key is identifying the dependencies. Using the dependency network with the generic activities helps make managing the project much easier.

Step 5—Estimate Effort

Many people classify estimating effort and time-scales as a finger in the air type of job—what some people call an educated guess. Some people say estimating accurately is impossible because we don't have enough information. Some people try to estimate at too high a level. Some people agree to a delivery date or give senior management a delivery date, just to keep senior management happy. They reckon that, as the project progresses, they will be able to point to all sorts of things that might be affecting or causing the committed date to slip. I've seen all this before, and at this point in the planning process, you have all the information you need to provide an accurate estimate for the phases you are planning.

The way you do it is to work with your team members, break each activity down to the tasks required to complete the activity and then estimate the effort and time-scales relating to the tasks. Add those task estimates together, and you have the estimate for the activity. Add the estimates and time-scales together for all activities required to produce a product, and you have the effort and time-scale for the product. Add together the effort and time-scales for all products in the phase, and you have the estimate for the effort and the time-scales for the phase plan.

OK, so it's not exactly that easy and, as the project manager, you must continually challenge the estimates for any extra contingency the Team introduces, but it's almost that easy to get accurate estimates of effort and time-scales. Follow the process. Breaking the activity down to its various, low-level tasks and then estimating the tasks provides you with a much more accurate estimate of effort than any other way of estimating.

Step 6—Identify Resource Requirements

Now that you know what must be delivered, the work to be accomplished in delivering it, the order in which the work must be done, and the estimate for effort to do the work, it is time to assign resources to each activity and task. It is best at this time in the process to know the resources by name. However, if you don't know their names because they haven't been assigned to you yet, then you can identify the resources by job function.

You could now find yourself in many situations. The first could be that you have been given a team of people to deliver this project. The second could be that you have been told to deliver by a specific date, and you must identify how many and who you want on your team. The third situation could be that you have been given a specific date and the team you have to deliver to by that date. The fourth situation could be.... There are many possible situations left. At any rate, now is the time to identify the number, skills, and type of team members you will need. I suggest that you not only identify the hard skills you want the people to have, but also the soft skills, behaviors, and attitudes.

If you have been given the people for your team, and they are not right for you, or there are not enough of them, or whatever other reason makes some, or all, of them unacceptable, or you have not been given any, now is the time to fight for them—the right people and the right number. Use the detailed project plan you are developing as the backup you need to prove your case.

Step 7—Prepare and Smooth the Schedule

The schedule is produced as an outcome of the effort required on an activity divided by the number of resources that can be and have been assigned to that activity. So it is not difficult to do by either hand or using an automated planning tool.

When I first moved to the United Kingdom from Virginia, I was responsible for putting together the plan for the computerization of the United Kingdom's income-tax system. The entire plan comprised about 7,700 activities for about 110 major products, with about 2,200 of those activities on the critical path. The 2,200 critical-path activities were a generic set of about 20 activities for each major product. Those 2,200 had to be scheduled. The other 5,000 were those that fit around the critical-path activities.

I remember a weekend at my house in Shrewsbury, England, when I removed the pictures from the living-room walls, put flip-chart paper on the walls around the entire room, and divided the flip-chart paper into weekly increments. I listed the 110 products down the left side of the first page, and started to schedule the project by hand, drawing the twenty-activity, generic dependency network for each product. It took two of us the entire weekend to complete the schedule, because once we drew the initial schedule, we had to smooth it to match the effort we had available from the planned resources.

Most automated planning systems advertise that they can be used to automatically smooth resources to the best use, thereby providing the best possible schedule. They say that they consider the activities, effort estimate, and resources allocated, by either name or resource indicator, and the estimated elapsed time for each activity. I have not found one yet that can do the smoothing as well as can be done by hand. Once you have smoothed the resources by hand, take the dates you end up with for each activity, put those dates into your system, and track achievement to them. It will be so much easier.

Multiple Uses for the Project Plan

I use the plan for many purposes, including the following:

- To determine how long a project will take.
- To monitor and control the project.
- To gain commitment of the team, bosses, peers, business, users.
- To help instill a teamwork culture in the project team.
- To get required resources.

- To identify how the required delivery date can be met.
- To fight arbitrary end-dates.

Using the Plan to Determine How Long the Project Will Take

This is an obvious reason. Once the plan is completed, everyone will be able to see how long the project will take.

Using the Plan to Monitor and Control

Once the plan is agreed, and all those involved in the project are happy with it, it should be used to monitor the project. The plan should be monitored at least weekly. I like to gather updates to the plan on a Friday, update the plan, reprint it, and distribute it on Monday. With the update information I receive on Friday, I can do whatever I need to do to control the project—move resources around; add resources, if required; question the dependencies, reschedule the activity, and so on.

Using the Plan to Gain Commitment

One of the most important things in the entire planning process is to have all stakeholders to the project committed to the project and the plan. It is especially important that project team members understand what they are scheduled to achieve, when they are scheduled to achieve it, who depends on their achieving it on the date scheduled, how that will affect the team member waiting for it to be achieved, and the consequences if the schedule is not met.

That commitment comes from understanding all this and accepting it. Once each team member knows all this, and she accepts the assigned activities, she is committed to it. Another thing that can be done to gain commitment is to have a walk-through review of the plan with each stakeholder or the stakeholder's representative and, at the end of that review, have him sign it to show that he understands and agrees it.

Using the Plan to Help Instill a Teamwork Culture in the Project Team

Developing the plan takes a lot of discussion and cooperation among the team members. I found that working together to create and agree the plan also helps each member feel part of a team. In addition, knowing, through an understanding of the plan, how each team member depends on each of the other members of the team helps instill the teamwork culture.

Using the Plan to Get the Required Resources

Once the plan is completed, and how many and what kind of resources are required the project manager can use the plan for a discussion with the resource providers. She can show how many and what kind of people are required to deliver her project successfully.

Using the Plan to Show How the Required End Date Can Be Met

Usually, when a project is assigned to a project manager, the project manager is also told when the project must be completed. There is then usually a big discussion about the end date, with the project manager finally giving in and agreeing to complete the project on the given date.

There is another way to approach it. The project manager should tell the customer that the date requested is definitely the indicated date for delivery. The project plan will be developed to show what is needed to meet the indicated date. Once the plan is completed, the project manager should meet with the customer and show what is needed for resources, staffing, and so on to meet the indicated date. If what is needed to meet the indicated date is available, then everything is great. If what is needed to meet the indicated date is unavailable, then it is up to the customer either to make the resources, staffing, and so on available or to change his requirement so it can be delivered by the indicated date or change the delivery date.

Using the Plan to Fight Arbitrary End Dates

A well-developed plan can be the basic information used to fight arbitrary end dates. That is, if you can't meet the arbitrary end date. If you can't have the resources the plan says that you need to meet the arbitrary end date, you must discuss this with your customer. If your customer continues to say that he wants it by that date, and you can't have the resources you need, then you must ask them one of the following questions (or maybe both):

1. What functions in the requirement do they want you to take out?

If the answer is none, then move to question 2.

2. What don't you want me to do in the development process?

If the answer is nothing, then you know you have an unreasonable customer.

Perhaps then, you must explain that you can only put five pounds, perhaps six pounds of stuff in a five-pound bag. Anymore, and the bag tears. Catastrophe! You can only put five liters of water, maybe a little more, in a five-liter bottle. Anymore and the water overflows the bottle. Catastrophe! Therefore, you can only give five people's work, maybe six people's, to five people, or you then have a catastrophe. The project slips, quality degrades, and you are constantly playing catch-up.

I've had customers continue to be unreasonable, and usually, I didn't take the contract. I understand it might be a little harder for you if you are an employee, but you need to let your customer know what you believe will happen and that you can't be committed to the arbitrary end-date plan.

Chapter 4: Points to Remember

Points to Remember

- Effective project planning and scheduling is essential for project success.
- Planning starts at the beginning of the project and is updated throughout the project until the project is completed.
- Schedule down to five-to ten-day activities.
- Establish new activities when responsibility for activity changes.
- Plans must be deliverable product-based.
- Activities must help deliver an agreed deliverable product.
- Have as many as possible of the project stakeholders involved in the production and agreement of the plan.
- Follow the 7-Step Planning Process as shown below to produce the project schedule.

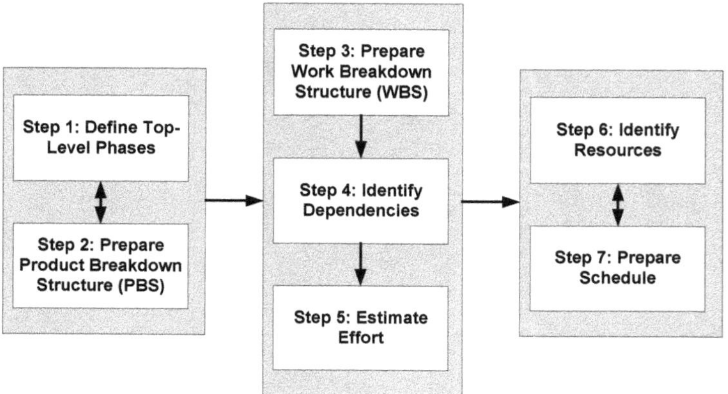

Use the project plan for many purposes, including the following:

- To determine how long the project will take.
- To monitor and control the project.
- To gain commitment of the stakeholders.
- To help instill a teamwork culture on the project.
- To get the required resources.
- To show how the required end date can be met.
- To fight arbitrary end dates.

Time to Do Some Work on Strategy 2—Plan the Achievement

Answer the following questions:

1. On a scale from 1 to 10, (1 being poor and 10 being so good that people throw rose petals on the ground in your path.) how well do you plan your projects?

____
____
____
____
____
____
____
____

2. Do you include project team members in the planning process?

____

3. Do you follow a standard planning process? If not, why not? Will you follow one?

____

____

____

____

____

4. Are all your plans deliverable-based? ___________

5. Do your plans go down to a five-to ten-day level of activity? ___________________________

6. Do you use your plan for things other than scheduling how long it will take to complete the job? If not, why not?

____

____

____

____

____

7. At what point in the planning process do you agree to a schedule and a budget? Do you have the detail you need to agree the budget and schedule at that time?

____

____

____

____

____

____

CHAPTER 5

Strategy 3—Organize for Success

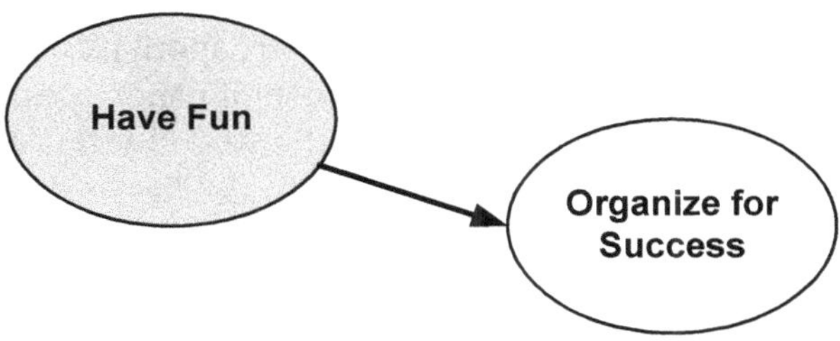

Fostering a Team Atmosphere on the Project

When the people on your project work together, they are much more productive, they have more fun, and they will deliver quality projects faster. How do you establish your project organization as a team? ManyWell, many characteristics must be met for the team concept to work. These characteristics include the following:

- **Common Purpose**—The team must have a common purpose and each team member must understand what that purpose is. As the project manager, you are responsible to ensure that the common purpose is communicated to be understood by the team members.

- **Goals and Objectives**—The goals and objectives for the team must not only be known and agreed by all team members, but the project manager must also ensure by actions and deeds that all team members share responsibility for the achievement of the goals and objectives.
- **Understand How Each of Them Fit In**—You must ensure that each team member understands the roles and responsibilities of not only himself but also of other team members. They must understand what each of them is expected to do to deliver success.
- **Achievement**—Achievement of the goals and objectives must be measured and made visible so team members will know exactly where the team stands with achievement.
- **Ground Rules**—The team needs to establish the ground rules that define how they will work together. These rules should also include how any conflicts will be handled.
- **Success**—The project manager should ensure that success is celebrated by the entire team, and the entire team is recognized as responsible for the success—some members in an active role and some in a support role.
- **Supportive Environment**—The project manager must ensure that the team operates in a supportive environment. Team members

should be able to be open without being concerned that they will be punished for their honesty.

I took over a project once that had spent about 70% of its budget and had delivered about 30% of its products. And those were delivered to questionable quality. The company was considering cancelling the project. The project staff was demoralized and couldn't see how they could succeed. After introducing the team concept into the project, being open with them about the situation, and asking for their help and ideas, they approached the project with renewed vigor and could agree and deliver to an approved rescue plan. Sure, we were a little late in delivering to the initial plan, but we were right on schedule delivering to the rescue plan. We were over the original budget but within the budget established as part of the rescue plan. And the quality of the delivered product was second to none. I attribute this turnaround to the introduction of the team concept. (And a great project manager, of course.)

The above characteristics are not the only ones that must be met for a team to be successful, but once these characteristics are met, they go a long way to helping the team get established and working.

If Possible, Co-Locate the Project Team

One thing I hate when I'm running a project is to have my project team spread all over the place. I like to have all my team in the same place. I'm well aware that this is sometimes difficult to make happen, but if possible, you should do what it takes to make it happen.

Instilling the soft skills into the team is much easier when the team members are in one place. It is easier for the team to be motivated, be more passionate about the project, be more committed to delivering a quality product, be more helpful with one another, share responsibility, and celebrate success when they are located together. It also provides a better environment for good communication within the team. If you remember in Chapter 1, poor communication was one a major reason projects are unsuccessful. So, the better the communication, the better chance the project has to be successful.

By the way, on IT projects, I'm not just talking about the IT people being co-located. I'm also talking about the client(s) assigned to the project. Now, I know that some of you are saying that is impossible. Well, when a business partner of mine took over a major project at a large insurance company, he was able to convince the client senior management to not only assign several client representatives to the project full time, but also to assign several others to test the system part time. In addition, when both the full-time and part-time user resources worked on the project, they were located with the project team, helping ensure that the product's final delivery was on time, within budget, and met the users' expectations.

Organizing to the Plan

Many times, I have looked at resource requests from my project managers and the organization of the resources, and there was little consistency among the resource requests, the organization, and the approved project plan. There was no consistency in numbers of people required at particular times and no consistency in how those resources would be used.

Too many times, a project manager will tell me that she needs, say, eight people for six months to deliver her project. When she tells me that, I'm sure that she is not re-sourcing to the plan because if she were, she would need fewer people at the beginning and at the end, with the resource numbers peaking in the middle. The plan must be the basis for the numbers of resources required and the time in the project when they will be required. The plan will help identify when the "right" people with the "right" skills will be needed.

In addition, when identifying resource requirements, we should consider not only the technical skills required, but also the soft skills we look for in our people. Finding the ideal people means finding people who meet both the technical and the soft skills requirements.

One other thing about organizing to the plan and getting the right resources and the right numbers—start negotiating with the owners of the resources when you have an idea of the people you will need. The sooner you start negotiating for them, the better chance you have of getting them when you need them.

Using a Project Board

It is common to establish a project board for each project. However, in many cases, the project board is not as effective as it could be for a number of reasons. For a project board to be effective, it must include, at least, the following members:

- The project sponsor.
- The manager of the group delivering the project.
- A person representing the users of the system to be delivered.
- If you have one, the person from the project support office providing support to the project manager and the project team.

In addition, I have the project manager report to the project board.

The project board members must be chosen for their commitment to the project. I've seen too many members of project boards be assigned to the board because they weren't too busy. I want members on my project board who will almost "die in a ditch" for the project. Well, maybe not actually 'die in a ditch' but I think you know what I mean. I want them committed to the project. Among other things, the project board members are responsible for the following:

- Approving the direction of the project.
- Approving the budget.
- Reviewing and approving the project documentation.

- Reviewing and approving changes.
- Ensuring that resources are available when required.
- Handling any major risks or issues that can only be handled at their level of the organization.
- Helping solve challenges the project manager faces, when required.

I have the project board meet at least monthly in a monthly achievement meeting where we report on and discuss achievement, any variance from the plan, and risks and issues. In addition, we review, agree, and approve project documentation; changes to project baselines; and we cover any other required business.

You need to remember this—a project board appointed to just "tick a box" is worse than having no project board at all! . With a tick-the-box project board, we have the false impression that we will have the help when we need it only to learn, too late, that they do nothing for us. At least without a project board, I know from the beginning that I will have to handle every challenge the project faces.

The Management Trio

I like to manage projects, especially large projects, using my 3 Amigos concept. The 3 Amigos are the project manager, the technical manager, and a project support officer. So what do these three people do? For what are they responsible?

The project manager is in charge of the project. She is the one responsible for making sure that the project is delivered on time, within budget, and meeting the clients' expectations. She is also responsible for ensuring that working on her project provides a positive experience for her project team.

Let me tell you why I appoint a technical manager as Amigo Number 2. Although I have successfully managed some huge technical projects in my life, I do not come from a technical background. I had many jobs until I started working on the Apollo Program as a messenger and then working my way up to managing the Change Control Board. My college degree is in business.

I'm not saying that I can't understand technical issues but I do not have the experience necessary to dig into the technical aspects of some projects. And, by the way, I am not that interested. In addition, I don't think that project managers have the time to dig deep into the technical issues, either.

When I was assigned to manage my first project, I took the assignment, though I didn't have a technical background, and the project was highly technical. So what did I do? I worked out a plan showing that I needed some hotshot technical people. They were assigned to me, and I made one my technical director. Her name was Rebecca, and she was just great. Her job was to spend her time leading the development activity and digging into any technical issues that arose. She would then brief me on the various choices associated with the issues.

My lack of technical experience didn't keep me from being intelligent enough to understand the problems. My job as project manager was to consider not only the technical issues but also the consequences around cost and schedule and make the decisions. It worked just great, and it was on that project, early in my career, that I realized that by appointing a technical director who understood the technical issues associated with the project, I could manage any project. A good project manager with a good technical director can take on almost every, if not every, type of project and deliver it. So the technical director's job is to spend time on the project's technical issues and brief the project manager on the various choices related to the project's technical elements.

The third person in the 3 Amigos is the project support officer who either manages a project support office (if the project is big enough) or *is* the project support office (if the project is not big enough to have a project support office team). This person is responsible for supporting the project team and me in a number of areas as shown below:

- Support of project planning, monitoring, reporting and control
- Configuration management
- Documentation management
- Action item management
- Risk and issue management
- Quality review management

Some organizations don't have a project support office and don't have project support officers. If that is so with your organization and your project, remember that the project manager must either take on the duties I have a project support officer doing or assign someone on the project to do it.

However, on my projects, I assign both a technical manager and a project support manager (2 of the 3 Amigos). Why? In my experience, I have learned that if I do everything a project manager should do in managing the stakeholders, motivating the team, managing the budget, reviewing the deliverables, ensuring that people stay focused, and so on, I won't have time to do the following things that should be done on a project. And, by the way, these are just a few things project managers should do.

- Look in detail at every technical issue.
- Prepare and document the project plan.
- Timely monitor achievement and maintain the project plan.
- Manage, control, administer, and report proposed changes.
- Identify, manage, and administer risks and issues.
- Establish and maintain a documentation library.
- Manage and administer the project documentation review process.
- Manage, report, and administer the action item process.
- Prepare and distribute reports on project achievement.

- Prepare for and attend achievement meetings, project board meetings, and so on.

I require help doing these things. Make no mistake about what I say here, I am responsible for these areas, but I take advice from my technical manager about technical concerns, and I use the project support officer's services to help me administer the management controls.

Many of you reading this book won't have the benefit of support from a project support office, but I'd like to spend a little time explaining what I have project support officers do. Remember, if you don't have a project support office, you must do these things.

The Project Support Office

The project support office is organized and implemented to do what it states—support the project, done by providing support to the project manager, the project team, and the project board. So what is the project support office, and how does it work? The project support office is a central pool of resources with particular support skills that can be resourced by one person on a small project to as many as fifteen to twenty people on big projects.

When a colleague was program director for a major High Street bank in the United Kingdom with more than five hundred people in his program working on about forty projects, he had a project support office of about twenty people. Depending on the size of the projects, a project support officer could be providing support to more than one project, or a project could have more than one project support officer.

Although you can make the project support office responsible for whatever you choose, at the risk of repeating myself, the common responsibilities are the following:

- Support of project planning, monitoring, reporting, and control
- Configuration management
- Documentation management
- Action item management
- Risk and issue management
- Quality review management

In supporting project planning, monitoring, reporting, and control, the project support office helps the project manager and the team prepare the detailed project plan, gather the data so management can monitor achievement to the plan, and prepare the various achievement reports required at all project levels. Make no mistake about it, though the project support office helps prepare the plan, the project manager and the project team own it.

In supporting configuration management on the project, the project support office is usually responsible for maintaining the records associated with each configuration item and configuration item baseline along with administering the change control process and maintaining a database associated with change control. The configuration item records include identifying the latest version of the baseline documentation, while the change control database includes the status of every change proposed to the various baselines.

In the action item area, the project support officer maintains a log of all action items the project manager assigns to project team members. It is his responsibility to monitor completion of those action items in support to the project manager and report on them at the achievement meetings. In documentation management, the project support officer is responsible for the implementation and operation of a project library, which becomes the repository for all project documentation, including the approved baseline documentation including approved changes.

In risk and issue management, the project support officer is responsible for implementing and operating the risk and issue management process, which includes setting up risk and issues workshops, maintaining the risks and issues database, gathering achievement information concerning the mitigation actions, and reporting on achievement in the various project reports and to the various management levels. The project support officer is responsible for supporting the quality review process, including distribution of items requiring review to applicable project team members, gathering of review comments to be presented at a quality review meeting, facilitating the review meeting, and maintaining the records required to support the process.

Working with the Users

Users must be a part of the project organization. I distinguish here between the project sponsor/business representative and the user of the item delivered. The person at the "coal face" will have a much better handle operationally than the project sponsor will. The project sponsor or business representative will have a much better handle on other areas and, therefore, he should be represented on the project.

The user, as a member of the project board, should be active in the development details, be part of testing, and should share the responsibility for delivering benefits with the project sponsor or business representative. The user activities should be included in the project plan and monitored the same as everyone else's on the project. In fact, anyone performing activities that affect the project's success should be included in the project plan.

Independent Acceptance

One of the most important parts of the project organization is an independent acceptance team. An independent acceptance team is set up to test the final delivered product. This acceptance uses the requirements document as its basis. It checks whether the delivered product meets its requirements. If the delivered product was delivered according to the requirements document, everything will be fine. If, however, changes were made without getting them assessed and approved, the product will not pass acceptance. So independent acceptance is testing what the product should do rather than what the product does. I hope that both those things will be the same, but I have learned that in too many cases, they aren't.

On an IT project, independent testing occurs after development testing when development can no longer access the system. It should occur before user testing begins, and only testers should resource it. The team should not include anyone who worked on the system's design or development. The requirements documentation should be the basis for preparing the independent test scripts, and the development team should correct any errors after a structured hand-over of the system back to development. Once the error has been corrected, the system should be handed back to independent testing using a tightly structured hand-over process. This tight hand-over process needs to ensure that changes made to the system were made only to correct the errors, and not to add or change other functions.

In short, the third strategy tells you to organize to the plan, establish a project board, use the 3 Amigos for greater success, establish and operate a project support office, include the users in the project, and establish an independent acceptance team.

Chapter 5: Points to Remember

Points to Remember

- Foster a team atmosphere on the project.
- Organize the project resources in line with the project plan.
- If possible, co-locate the project team including the users assigned to the project.
- Establish a project board to support the project and the project manager.
- Organize management into the management trio to include:
    - The project manager
    - The technical manager
    - The project support office manager
- Establish a project support office to help the project and the project manager at least in
    - Project planning, monitoring, and control
    - Configuration management including change control
    - Risk and issue management
    - Documentation management
    - Quality review
    - Action item management
- Work as closely as possible with the users

- Establish an independent acceptance team on the project.

Time to Do Some Work on Organize for Success

Answer the following questions.

1. Do you foster a team atmosphere on your project?

___

2. Do you organize your resource requirements in line with the project plan? ___

3. Is your entire project team in the same area? If not, how do you keep the team working together? Are the client representatives co-located with the development team?

___

___

___

___

___

4. Do you work with a project board or something similar? Who is represented on the board? If you don't have a project board, why not?

___

___

___

___

___

___

___

5. Do you think that the 3 Amigos concept has real benefit for you? Would it make sense on your project? Who would be your other 2 Amigos?

6. Do you have a project support office? Would a project support office be worthwhile on your project? If you don't have a project support office, how do you find time to do the following:

- Project planning, monitoring, and control
- Configuration management, including change control
- Risk and issue management
- Documentation management
- Quality review
- Action item management

7. What is your relationship with the users and/or the clients?

8. Do you use an independent acceptance team? If not, why? How do you ensure that what you hand over to the client meets his or her requirements?

CHAPTER **6**

Strategy 4—Gain and Maintain Commitment

Importance of Commitment

Commitment is essential for project success. And who do we need to be committed? Every stakeholder on the project. That means you, the project manager, project team members, customers, users, other project managers whose project interfaces with your project, the project sponsor, the project board, and whomever I forgot. I want everyone who has a share in project success to be committed.

Many times, people say they are committed, but they are only involved. What's the difference you ask? Well, I find it difficult to define exactly what commitment is. The online dictionary defines *involvement* as "connected by participation or association or use," while it defines *commitment* as "the act of binding yourself (intellectually or emotionally) to a course of action." However, let's look at an example that shows the difference between commitment and involvement well. We have to look no further than your unhealthy breakfast this morning. If you had a breakfast of bacon and eggs, you should note that the chicken was involved, but the pig was committed.

What Is Real Commitment?

Real commitment on a project is difficult to define, something you feel before you see it. You feel this buzz when you have anything to do with the project—the good vibes you feel when you talk to team members about the project or when you walk into the project room. You feel it, and then you see it. Have you ever worked on or managed a project with that kind of buzz. I have, and I tell you it's a wonderful feeling. Here's hoping that if you haven't already, you will experience one soon—because it feels great.

What Made the Project a Buzz Project?

There are many criteria that make a project a "buzz" project, including the following:

- People-oriented project manager.
- Teamwork.
- Team knows and owns the outcome.
- Tight but achievable schedule.
- The team feels important.

Let's look at each criterion in detail.

People-Oriented Project Manager

One key thing that made the project a buzz project was that a people-oriented project manager ran it. He cared about his people, and he was fair in all his dealings with them. He asked questions and valued their input. He understood that he was there as a servant to the team and always looked for ways he could make their job easier.

He was a project manager who understood the hard skills tools, techniques, and processes, at the same time, practiced the soft skills related to his attitude and behavior. He was a positive person and always looked for the best in people and the opportunity in all the challenges the project faced. He was a "work hard, play hard" person who was easy to be around, and the people on the project could relate to him. In general, he was an all-around good person with excellent interpersonal skills.

Teamwork

The project manager fostered a team atmosphere on the project. Those working on the project felt as if they were part of a team. What did he do to introduce teamwork on the project? He made sure that the entire team understood the project's common purpose. He made sure that the goals and objectives of the project were known and agreed by each team member. Each team member knew how he or she fit into the project. Each team member understood his or her role and responsibilities and the roles and responsibilities of other team members. It was understood what each was expected to do to deliver success.

Achievement of the team goals and objectives were measured and made visible so team members knew exactly where the team stood with achievement. Ground rules how the team would work together were established, including how any conflicts would be handled. The project manager made sure that the entire team celebrated achievement. He made sure that the entire team was recognized as responsible for the success—some members in an active role and some in a supporting role. The project manager ensured that the team operated in a supportive environment. Team members could be open without being concerned that they would be punished for their honesty.

The Team Knew and Owned the Outcome

The team was well aware of what the goal was—that is, what was supposed to be delivered. A requirements document was in place and under control. They read and understood the document and the process and procedure for controlling it. Not only did they understand what they were delivering for the requirement, but they also had a clear idea of the benefits to the business that they were delivering.

Tight but Achievable Schedule

The project schedule was tight but achievable. This tight schedule caused the people on the team to work with a sense of urgency. I have always found that if you want something done, give it to a busy person. Just as she is making sure that her job is done, she will make sure that the task you give her will be done. Schedules that are not tight cause the team to start becoming slow at what they do, and before you know it, the project with the loose schedule starts slipping, while the project with the tight schedule continues merrily on its way right on schedule, that is, if a great project manager manages it.

The Team Felt Important

The team felt important because they were making changes or delivering a new system important to the business' success. Because of their treatment, they knew that their part, which might be considered small in the overall project, was required by all other parts to deliver the system, changes, or whatever they delivered, successfully. For instance, during the Apollo Program, one astronaut asked a janitor sweeping the floor in the Vertical Assembly Building at Cape Canaveral what he was doing. His answer wasn't sweeping the floor. His answer was helping put a man on the moon. That's what I mean about owning the outcome and feeling important.

Let's Look at No-Buzz Projects

What causes the project to be a no-buzz project? There are loads of reasons, but I want to write about the ones I have run into in my career and the ones I believe are the major ones. These include the following:

- People kept in the dark
- Poor leadership
- No teamwork—every person for himself
- Outcome not well known or controlled

People Kept in the Dark

Have you ever worked on a project where communication was terrible, where you were kept in the dark about things? Things happened that affected you but because you didn't know exactly what they were, you couldn't do anything about them. I have, and I must tell you that it is a horrible way to work. You want to know what's going on because you know you can help, but they, the authorities, won't communicate the problem, the issues, or the situation they face. It starts making you feel as if they don't value your input or that you are not important enough to know what's going on. You know things are happening that you should know about, but they won't tell you. This won't make you feel "buzzy" or part of the project.

Poor Leadership

Having a person on the project with a title of leader who isn't a leader or is a poor one is another reason for no-buzz projects. Leaders need to take charge while realizing that they are there to serve their team members. Bad leaders don't look at leadership this way. They either see themselves as dictatorial types, taking control, in charge, telling people what to do and how to do it. They get the title of leader, and they go crazy. Or, they see themselves as the nice person who wants to be the team's friend and, therefore, doesn't want to do anything to upset the team.

Every Person for Himself

Project teams where people, including the project manager, are just out to make themselves look good, even at the expense of other team members is another cause of a no-buzz project. Have you ever worked on a project where someone else claimed the idea as his or hers when you devised a great idea? Or, you achieved something difficult to do, and the boss took credit for it. Early in my career, I worked on several projects with people like that, and it didn't make me happy. Talk about no-buzz!

Let me give you an example. Many years ago, I worked for a boss I believe would have won the gold medal at the Olympics if there was a competition for the worst boss. Everything you had to do as a leader to motivate your people, he did the opposite. He didn't motivate me, and he was a big problem to the project manager running one of his projects.

He assigned me to direct the project. The project was far behind schedule; the team's morale was poor; there was no plan in place; the team was not motivated in the least; and the team faced many other challenges. The project had used about 3/4 of its budget, and the team had completed about 1/4 of the work. I worked hard with that team to raise morale, to put an acceptable plan in place, to organize the team so they were more productive, and to put a better environment in place—an environment where their natural motivation could come through.

They worked hard, and they were committed to bringing the project in on time with the new plan. I couldn't help them do any technical stuff when they worked on Saturdays, but I could fix a good old New Orleans dish—red beans and rice or gumbo—and bring it in and serve them. I supported them in whatever way I could. They were highly motivated. And we were successful in delivering a quality project in line with the new plan and that met the client's expectations.

A week or so later, I heard from a friend who had been in a senior management meeting with my boss, other senior managers, and the company's CEO. He told me that my boss had reported that he had successfully delivered the project—the project he gave me because, and he would never admit this, he was screwing it up. The CEO praised him for his success, and he never mentioned my name or the names of any team members. I was angry. I went to his office to confront him with this information and to ask him why he did that. He then had the nerve to say, well, I motivated you, and you just motivated the people on the project so I consider that I should get any accolades for delivery.

Outcome Not Well Known or Controlled

Another reason for a no-buzz project is that the requirement for what the team was delivering was not well known. The project might have started with everybody thinking they knew what the outcome was. I say "thinking they knew" because, usually, on no-buzz projects, the project lacks a good requirements document. So different people think the requirement is different things. Then, changes are made without having them under control and because of poor communication (another reason for no-buzz projects), not everyone hears of the changes, and chaos ensues. Part of the no-buzz project is that the team, because they didn't know what the finished project would deliver, also had little idea of the benefits the project was delivering to the client.

You have probably heard the story of two bricklayers working on a beautiful cathedral. A stranger walked up to them and asked what they were doing. The first said he was just laying bricks. The second said that he was building a cathedral. The second had studied the plans, looked at the rendition of what the cathedral was going to look like, and in so doing, was motivated working on this cathedral. He got a buzz out of what he did. He knew that this beautiful cathedral would be used as a house of prayer, and it would probably last for centuries. He saw himself as part of the bigger picture. The first just saw what he did as a job.

To have a buzz project, all team members should know the outcome concerning the big picture and the benefits their project delivers.

How to Gain Commitment

I started this chapter by saying how important commitment was to the project's success. I also mentioned that I thought all stakeholders needed to be committed. One way to help make this happen is through the Committed/Not Committed Matrix. An example of this matrix is shown below:

Committed/Not Committed Matrix			
Stakeholder Name	**Position on Project**	**C/NC**	**Action to Be Taken**
John Doe	Project Manager	C	None
Jane Doe	Sponsor	NC	Continue to stress her importance to the project/keep her informed
A. N. Other	Operations Manager	NC	Meet with her Invite her to meetings. Ask her for early input. Keep her up to date
Jimmy Doe	Test Manager	C	Ensure she continues to feel important to the project

This matrix provides the project manager with the visibility and the action planning required handling commitment on his project. What are some actions the project manager can take to gain uncommitted stakeholders' commitment? You've seen some in the preceding example, but let's look at several actions that can be taken in detail.

Get People in the Organization Involved

That includes the bosses, the team, your peers, and all stakeholders. Invite them to your project meetings. Present to them about the project and the benefits to the organization when the project is delivered. Keep them up to date on project achievement.

Foster Teamwork within the Project Team

Many things must be in place for the project team to feel like a team. I've written about teamwork in various places in this book, and I don't mind mentioning it again. It is one of the most important things project managers must do to deliver successful projects.

- **Common Purpose**—The team must have a common purpose, and they must understand what that purpose is. As the project manager, you are responsible to ensure that the common purpose is communicated to be understood by all the team members.
- **Goals and Objectives**—The goals and objectives for the team must not only be known and agreed by all team members, but the project manager must also ensure by actions and deeds that all team members share responsibility for the achievement of the goals and objectives.
- **Understand How Each Fits In**—You must ensure that each team member understands the roles and responsibilities of not only himself, but also all other team members.

They must understand what each is expected to do to deliver success.

- **Achievement**—Achievement of the goals and objectives must be measured and made visible so all team members know exactly where the team stands with achievement of the plan.
- **Ground Rules**—The team needs to establish the ground rules that define how the team will work together. These rules should also include how any conflicts are handled.
- **Success**—The project manager should ensure that the entire team celebrates success, and the entire team is recognized as responsible for the success—some members in an active role and some in a support role.
- **Supportive Environment**—The project manager must ensure that the team operates in a supportive environment. Team members should be able to be open without concern that they will be punished for their honesty.
- **Solicit People's Help and Backing**—Anytime you need help in getting things done that you can't handle yourself, ask for help from stakeholders. People will help and will feel good that you thought of them to help.

Ensure Stakeholders Know Their Importance to the Project

You can do this by referring to anything they might have done for you and the project. Thank them for their help on the project, and while doing that, mention how important they are to your success. Ask for stakeholders' advice when you face a challenge on which you would like some help. Ask them to review project documents, processes, procedures, and so on. Show how much you appreciate their input.

Communicate, Communicate, and Most Important, Communicate

You can communicate in many ways, including the following:

- **Project Newsletter**—Put out a project newsletter with information about achievements on the project and about the people on the project team. For instance, perhaps the client approved your project design. Article in the newsletter. Joe on the project team is marrying. Article in the newsletter. Mary had a baby. Article in the newsletter. Make it interesting.
- **Briefing Sessions**—Invite stakeholders not on the project team to briefing sessions about the project. Include things such as where you are on the project with schedule and budget. What major activities are coming up? Remind them of the benefits the client will accrue when the project is delivered. Get the client to brief the project team on preparation in the client's facility for the delivered project.
- **Visibility Room**—Use a meeting room's walls. Put the latest information about the project on the wall—things such as the project's name and a description, the project manager's name, the team members' names,

the deliverables to be created and delivered, the project schedule, project achievement so far, and whether the project is on schedule.

Respect the individual.

We project managers must remember that, though we stress the team concept, we still need to respect each person on the team. Ask for their opinions and listen to them. We need to put our interpersonal skills to good use and remember to treat each person separately.

Get the appropriate people to sign.

Getting people to sign is a good way to get commitment. I ask various stakeholders to sign various documents. For instance, getting the client to sign the requirements shows that he agrees with what is in the document for what needs to be delivered. Getting the team to sign the requirements means that each team member is committed to delivering what the client said he wanted. I like the plans to be signed by team members and, occasionally, even the client. In signing the plans, I want to make sure that those who sign understand and are committed to their activities and the schedule for completion of each of those activities. I've even thought, although I've never done it, about getting a team member to sign his or her work with the statement that "this is the best I can do." I might do it yet and see what happens.

Make sure people enjoy themselves.

I don't mean that you should treat work as an amusement park. What you are delivering to the client is serious, but you don't have to take yourself so seriously. Lighten up. Smile. As the song goes, "don't worry; be happy." Sponsor socials. Work hard and play hard. Enjoy yourself, and make it enjoyable for the team.

How to Maintain Commitment

Maintaining commitment is simple, but not necessarily easy. To maintain commitment, you need to keep doing all the things you had to do to gain commitment, plus one more thing—**deliver**.

Chapter 6: Points to Remember

Points to Remember

- Many successful project managers I have interviewed have made the point that commitment is essential to success.
- The dictionary defines *commitment* as "the act of binding yourself (intellectually or emotionally) to a course of action."
- You know commitment when you feel a buzz when you walk into a room filled with people committed to their project.
- Some conditions that cause the buzz include the following:
    - People-oriented project manager
    - Teamwork
    - Team knows and owns the outcome
    - Tight but achievable schedule
    - The team feels important
- No-buzz projects are those without committed teams working the project.
- Some conditions that cause no-buzz projects include the following:
    - People kept in the dark
    - Poor leadership
    - No teamwork—every person for himself
    - Outcome not well known or controlled

- Some ways to gain commitment include the following:
    - Use of the Committed/Not Committed Matrix.
    - Get people in the organization involved.
    - Foster Teamwork within the Project Team.
    - Ensure Stakeholders know their importance to the Project.
- Communicate, communicate, communicate in many ways, including the following:
    - Publish an informal project newsletter
    - Facilitate briefing sessions
    - Establish a visibility room
    - Respect the person
    - Get the appropriate people to sign
- Ensure that people enjoy themselves.
- Maintain commitment by continuing to do everything you did to gain commitment, plus—deliver.

Time to Do Some Work on Gain and Maintain Commitment

Answer these questions.

1. Are all your project stakeholders committed?

2. Do you run "buzz" projects? To what do you attribute your buzz project? People-oriented project manager? Team working together? Team knows and owns the outcome? Tight but achievable schedule? The team feels important? Or other reason?

3. Are any of these criteria in place if your project isn't buzzing? People kept in the dark? Poor leadership? No teamwork—everyone out for himself or herself? Outcome not well known or controlled? Or other reason?

4. Is there anyone on your project whom you think is uncommitted to the project? Which way to gain commitment will you use to get him or her on your side?

5. What is communication like on your project? If it could be better (and let me say that most communication on projects could be better), what will you do to make it better?

6. Deliver!

Chapter 7

Strategy 5—Take Massive Action

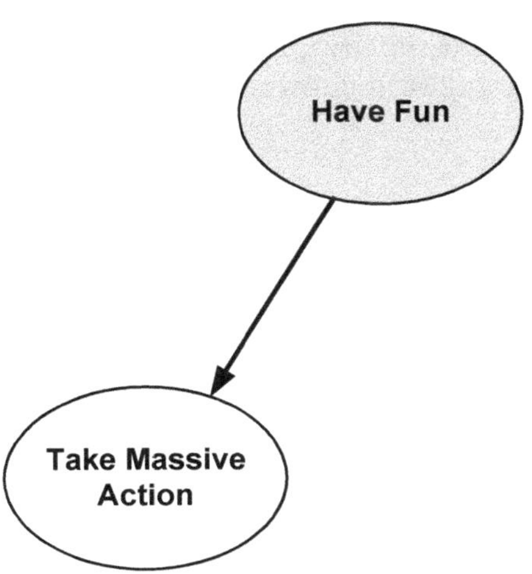

Action Does Not Equal Achievement

How many times have you finished the day and felt as if you had not achieved anything. Oh, you've had enough action during the day. You've been busy but... If you are like most people, you probably have spent days where you have been busy and not achieved anything worthwhile.

I had a contract with a large computer company in Europe, and I needed to speak to a director. It was well known in the company that he was surrounded by of action but very little achievement. I talked with his secretary and told her that I needed just fifteen minutes of his time. She told me that he didn't have a free fifteen minutes for more than a week. She said he was so busy that she couldn't talk with him about my meeting request for a few days. Sometimes, you just have to use a little humor. I asked her when he took a shower because I'd take a shower with him if it meant I could see him in the next day or so. I broke through. She laughed and said that he would see me if I showed up at his office at 7:15 the next morning. I did, and it looked as if he had already taken a shower. Thank goodness!

Action Equals Achievement

The action that equals achievement is action to the agreed plan. When the team starts to follow the plan with urgency, that is taking massive action. If, as the project manager, you find that a team member is doing something not in the plan, you need to look at it from two ways. If the activity must be done, it should be added to the plan. If the activity doesn't have to be done, then get them to stop doing it.

You and the team members should always work to the plan, and if things aren't working as you thought they would work, then change the plan. It's your plan.

How to Take Massive Action

The first thing one must do to take massive action is to know what action to take. And where does that come from? It comes from the project plan and schedule of activities. The next thing for the team member to do is to accomplish what they should, according to the plan, with 'sense of urgency. That sense of urgency is important because it is the foundation for taking massive action. So, what is the best way to ensure that you are doing what you should be doing when you should be doing it? Know the plan. Know how you fit into the plan. Prioritize your day so you will do the things you should be doing today, and not just what comes up when you get to work.

The To-Do List

How many of you use a to-do list to plan your day? I mean you make a list of all the things you must do, you start doing them, and as you finish, you scratch through the item on the list. Now, I will ask you something that could be embarrassing. If you do something not on your list, do you add it to your list and immediately scratch it off? Come on. I know most of you do that. Why? We all know it feels good to look down your list and see all those things crossed out.

Most people take a shot at prioritizing the items on the list using the "gut feeling" method. But at some point during the day, they change their mind; they have new things added to their list; they jump from one thing to the other; and at the end of the day, they have more on their list than when they started. They have things on their list that were probably more important and urgent than things they completed. Let me show you another more structured way to prioritize, and I guarantee that if you use this structured method of prioritization, you will be up to 35% more productive during your day.

The Prioritization Process

My suggestion is for you to prioritize your list items by how important the item is and how urgent it is. This will help you in your prioritization process by giving you some organized way to prioritize rather than by the gut feeling method. Important activities are those activities that take you closer to your goals and your situation at work. For the project plan, they are the activities needing to be done to stay on schedule. Urgent activities are those activities that must be done now. This urgency usually comes from some requirement outside you.

Step 1—Make Your To-Do List

You should go ahead as you normally do and create your to-do list. That is, list everything you believe you need to do. List them in no particular order and don't try to prioritize them at this time. Just make the list.

Step 2—Classify Items

Going down your to-do list, mark each item as either Urgent or Not Urgent. In addition, mark each item as either Important or Not Important. Once that is done, you have a to-do list on which each item is classified by its importance and urgency. Each item is classified in one of four possible conditions as follows:

- Condition 1—Important/Urgent
- Condition 2—Important/Not Urgent
- Condition 3—Not Important/Urgent
- Condition 4—Not Important/Not Urgent

Let me explain how the typical person relates to the various conditions. First, when you work on items included in Condition 1, the Important and Urgent Condition, you are reactive. That is, you react to events around you. It's as if someone else is "pulling your chain." If you have items in this condition, you should work to get rid of them as quickly as you can and then start spending your time on Condition 2 items.

When you work on Condition 2 items, that is, items that are important but not urgent, you are proactive. Your goal should be to work on these Condition 2 items. Working on Condition 2 items puts you in control. You have unhurried time to do the important things on your list. You can do better-quality work. You will have less stress in your life. These Condition 2 items should be items related to meeting your goals.

Condition 3 items are the items that most people start working on when they finish their Condition 1 items. They are urgent, but they are not important for meeting your goals. They might be important in terms of someone else meeting his or her goals but not yours. Because of this, you must decide whether you should spend time doing them. And I wonder why anyone would do Condition 4 items.

Step 3—Pick the Top 6 Items You Want to Achieve Tomorrow

Now that you have you items on your to-do list prioritized by importance and urgency, pick the top six that you want to do tomorrow. They are your top six, and you will use your criteria in selecting the top six; however, I suggest that you spread your picks between Condition 1 and Condition 2 items. The idea is to clear up the Condition 1 items—items causing you to react as quickly as you can—while not letting items slip from Condition 2 to Condition 1. You want to get out of this reactive mode.

Step 4—Prioritize Your Top Six Items

You now must prioritize the top six items you chose as wanting to do tomorrow. Prioritize them from 1 to 6. Now, close your planning tool and forget it until tomorrow. Enjoy today. Enjoy your evening. Your subconscious mind will start working on the list, and tomorrow, it will provide you with thoughts and ideas about how to tackle your top six.

Step 5—Attack Your Top Six Items in Prioritized Order

Tomorrow morning, when you start work, clear your desk of all distractions and start working on Number 1 on your prioritized list and don't stop working on Item Number 1 until you complete it. You might be interrupted. Your boss might stop by and ask you to do something for him; your phone might ring with a friend requesting your help on something.

Now is the time you must make a decision. You need to ask yourself, "Is what is asked of me a higher priority than Number 1 of my six selected items?" If you decide it is, then do it, but when you get back to your desk, keep working on Number 1. And work on Number 1 until you finish Number 1. Then, move on to Number 2 and so on. You might only get two items completed out of your six, but remember they are the top six items you picked from all the items you started with on your list. And you picked them out by using a structured approach to your planning, not the old gut feeling approach.

This process makes you more productive, but it only makes you more productive if you do it. Just knowing about it doesn't help your productivity.

Knowledge = Power

You probably know the saying that "knowledge equals power." I don't believe that. I believe that taking action is power. Why do they call the person who points out that the river is dirty an activist when, in fact, the person who cleans it up is taking action. Shouldn't the person taking the action be called the activist?

Chapter 7: Points to Remember

Points to Remember

- Action does not equal achievement.
- Planned action equals achievement.
- Incorporate the structured prioritization process into your daily planning process:
    - Step 1—Make your to-do list.
    - Step 2—Classify each item by importance and urgency.
    - Step 3—Pick the top six items you want to achieve tomorrow.
    - Step 4—Prioritize your top six items.
    - Step 5—Attack your top six items in prioritized order.
    - Clear desk of all distractions.
    - Complete one item before moving on to the next.
    - Requests during the day need to be prioritized against your top six.
    - Always return, after distraction, to the item you were working on before the distraction.
- Just knowing something doesn't give you the power. Taking action on what you know gives you the power.

Time to Do Some Work on Take Massive Action

Answer these questions.

1. Is your project committed to taking massive action to the plan? If not, why not, and what will you do to get them committed?

____
____
____
____
____
____

2. Do you seem busy during the day, but at the end of the day, you wonder what you achieved?

____
____
____
____
____
____

3. Do you prepare a to-do list daily? Would it be better to adopt the structured prioritization process in your daily planning? If not, what will you do to stay proactive in your daily activities?

____
____
____
____
____
____

CHAPTER 8

Strategy 6—Monitor and Control Achievement

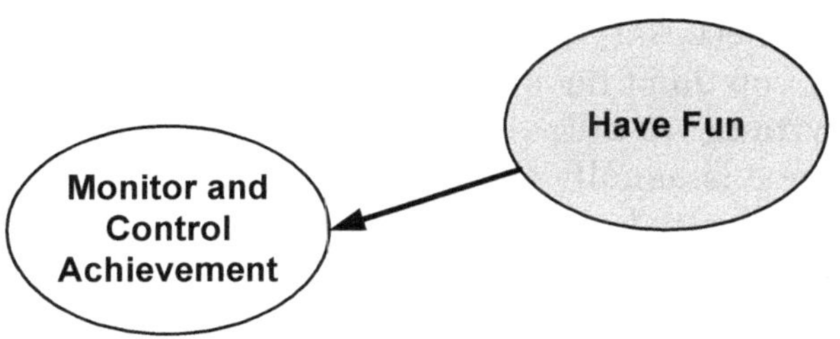

Monitor and Control Achievement

Once you have your requirement defined, your plan in place, your team organized, the right people committed, and people have started to take massive action, it is now time to monitor and control achievement to your plan. In addition, the project manager must use many disciplines to keep his project within budget, on schedule, and meeting the client's expectations. These disciplines include the following:

- Achievement planning, monitoring, and reporting.
- Change control.
- Risk and issues management.
- Documentation management.
- Visibility room/wall.

Achievement Planning, Monitoring, and Reporting

One important thing for the reader to understand is that I use the project support office to support the project in monitoring and controlling the project. It is unnecessary to use a project support office, but if you don't have one, then someone must do everything we talk about in this chapter, and that someone is usually the project manager. However, I've generally learned that the project manager does not usually have the time to do all this planning, monitoring, reporting, and controlling, and some is left undone. Not so good for the project, which is why I use a project support office.

Achievement Planning

We've covered project planning in detail in Chapter 3, so it's no use trying to cover that in any more detail. However, for achievement monitoring, I would like to write about several things.

Achievement Monitoring

Having a great achievement monitoring process in place on your project is essential to project success. The project manager should design, develop, and implement an achievement monitoring process including steps similar to the following:

Step	Action
1	Team members report on achievement or non-achievement of their scheduled activities.
2	Project support office/project manager collects the updates and includes them in a preliminary updated plan.
3	Project support officer reviews updated preliminary plan with project manager (unnecessary step if no project support officer).
4	Project manager discusses any negative variances with the applicable team member and decides on the action to take to correct situation.
5	Project manager approves updated plan.
6	Project support officer/project manager makes approved updated plan available to team.

Achievement Reporting

Achievement reporting is one thing that usually takes up much of the project manager's time, and I don't think it has to. At the beginning of the project, the project manager should define, in consultation with stakeholders, a complete list of the achievement reports required during the project. Templates of the required reports should be prepared and the templates should contain the format and content standards for the required reports. In addition, whenever a request for a new achievement report is made of the project manager, the project manager should see if any existing reports being prepared would suffice. If no existing report meets the requirement, a format and content standard for the new report should also be prepared and agreed with the requestor.

As much as possible, you should try to make the report come directly from the project plan, showing for the period covered:

1. What should have been achieved.
2. What was achieved.
3. What is scheduled to be achieved during the next reporting period.
4. An explanation of variances from the plan.
5. A presentation of the way forward.
6. A discussion of challenges the project faces.
7. A way to handle the challenges.

Most achievement reports (most project managers call them progress reports or highlight reports) are too wordy and do not provide easy-to-understand achievement information. I've also discovered that as a project gets into trouble, the reports tend to go one of two ways. Either they get wordier and wordier, or they are rare, with excuses about why they are late, and so forth.

Change Control

Let's look at how to control changes on your projects. Let's start by defining a change. A change is nothing more than a variation from an agreed position.

Project changes are probably the project manager and the project's biggest threat. Almost everyone has heard of scope creep where the requirement's scope keeps changing. I don't mind scope creep if it is under control. And I'll show you the kind of control I mean a little later. Most people see change on the project as a bad thing, whereas not all changes are bad. For instance, during system design, it might be proposed that if the requirement were changed just slightly, the system could be built less costly. If everything on the change were acceptable, that would be considered a good change. By doing it, we could save the client some money.

I have never asked project managers if they control changes and have them say, "No, we just let those suckers fly. We don't control them." Every project manager I have asked that question of has said, "Yes, we absolutely control changes on our project." However, I have seldom seen projects where changes were controlled properly. There are many reasons for this, but the chief one is that the project manager generally doesn't have the time to control them as rigorously as they should be.

Anyway, the project manager should design, develop, and implement the project change control process. Change control is not a difficult procedure to put in place, but it is sometimes difficult to get people to follow it. We need to be creative when designing and developing the change control procedure and bureaucratic in following it. What usually happens is that the change control procedure is usually bureaucratic, and people are creative in following it.

The basic process, as with so many processes, can be shown in a relatively simple manner.

Step	Action
0	The process starts with an approved "base-lined" document and someone on the project who wants to change it somehow.
1	The person proposing the change describes the proposed change on a change request, a change proposal, or a request for change (call it whatever you want) and submits it to the project support office/project manager.
2	The project support officer/project manager updates a change log and distributes the change request to the applicable team members for assessment.
3	The team members comment on the cost, schedule, and technical impact of the proposed change and send the assessment to the project support officer/project manager.
4	The project support officer/project manager chairs a change control board meeting where the proposed change is discussed, and the appropriate member of the board-the person who controls the baseline being changed either approves, rejects, or defers the pro-posed change to allow the gathering of more data.
4	The team implements the decision.

As anyone can see, this or a process like this can easily be designed, developed, and implemented on the project.

Documentation Management

Most projects do not have a project library where copies of documentation are kept. Project documentation is usually handled in a somewhat slipshod fashion. Project documentation is handled properly, usually because the project manager recognizes that it must be done, not because the organization has project library standards in place.

You, the project manager, should design, develop, establish, and implement your project library (but remember, if you had a project support officer, she could do it for you). You should prepare procedures required to run the library, including processes for storing the project documentation, processes for reviewing and approving documentation, and processes for checking documentation in and out of the library. The following stepped process is a sample of what should be developed on your project to manage documentation.

Step	Action
1	Team member(s) produce a draft of a project deliverable and submit it to the project support officer/ project manager.
2	The project support officer/project manager includes the draft of the deliverable in the project library and distributes the draft to applicable project members for review and comments.
3	The project support officer/project manager chairs a meeting to discuss and approve selected comments.
4	The document's writer incorporates the approved comments and submits the final deliverable to the project support office/project manager.
5	The project support officer/project manager includes the deliverable in the project library; it is put under change control; and changes to the deliverable must follow the change control process.

Risk and Issues Management

Risk and issues management is one of those things we know we need to do, but often, we don't seem to have the time to do it well. So, what do many project managers do? They give lip service to it. They put a risk and issues list together, and now and then, they update it, but between the updates, they don't have the time to manage the list.

As an aside, another thing I find is that there is usually a big discussion of whether something is a risk or an issue. From a project management standpoint, who cares? Some action will have to be taken whether it is a risk or an issue, and you just need to make sure that the action is taken. How do I do it to make sure that it gets done?

As the project manager, you are responsible for defining, developing, implementing, operating, and administering the risk and issues management system. The following is an example of risk and issue management activities that need to be part of the management process:

1. Initially identifying the project's risks and issues, which is usually done in a workshop attended by applicable stakeholders. If you don't have a project support office, as project manager, you must facilitate the workshop yourself, or sometimes, the company has professional facilitators on whom you can call.
2. The analysis of each risk and issue. Each risk is analyzed for its impact and the likelihood of its happening. Each issue is analyzed for its impact on the project if not taken care of and its urgency. I like to use a scale of 1 to

4 in both evaluations with 1 being low and 4 being high. Both numbers on each risk and issue are multiplied, and you come out with a single score. You can decide at which score level you will pay attention to the risk or issue. I pay particular attention to any risk or issue that scores 8 or higher.

3. A discussion and agreement of what actions must be taken to mitigate the risk and handle the issue, who will be responsible for taking actions, and by when the actions must be taken.
4. Monitoring of the actions to ensure that the actions are taken on time.
5. Reporting on the achievement of the plan to mitigate the risks and to handle the issues.
6. Continued identification, analysis, and so on of risks and issues during the life of the project.

Visibility Room

One of the ways to keep the project team, stakeholders, management, and any other interested party up to date with the project's achievements is through the use of a Visibility Room. The walls in the visibility room are used to display information about the project, or about all of the projects being managed in the organization. The type of information we have included on the walls include the project name, the project summary, name of the project manager, names of the project team members, an organization diagram, project schedule down to the deliverables level, the team member responsible for each of the deliverables, achievement information on each of the deliverables and any other information the stakeholders would like displayed.

Usually we used a meeting room which could continue to be a meeting room because all we needed were the walls. The meeting room was a good place to have the information because it would be visible to a large number of members of the organization - every person that attended a meeting in the room. If a room is not available, then a wall in the hallway where people can see the information will work just as well.

There is some effort required to keep the information up to date and I had a Project Support Office on every project where I had a visibility room. In addition, the support office was assigned the responsibility of maintaining the room.

I, also, liked to use two kinds of stickers to put next to the deliverables. I used a 'blue angel' sticker if the preparation of the deliverable was on schedule and a 'red devil' sticker if the preparation of the deliverable was late. Some organizations did not like the idea of the stickers as putting a red devil next to a deliverable would be like we were blaming the person responsible for the deliverable for being behind schedule. Personally, I liked the stickers as I thought they added a little more motivation for the person responsible for the deliverable to get it in on time.

Chapter 8: Points to Remember

Points to Remember

Many monitoring and control disciplines help keep the project on track. If possible, set up a project support office to support the project manager and the project team in the following monitoring and control disciplines:

- o Achievement planning, monitoring, and reporting
- o Change control
- o Risk and issues management
- o Documentation management

- Having a great achievement monitoring process in place on your project is essential to project success.
- Achievement from the team should be reported weekly.
- The project manager should verify all reported achievement.
- The project plan should be updated with the verified achievement and distributed to the various stakeholders.
- Achievement reports should come directly from the project plan with variances from the project plan explained.
- Scope creep happens because of a lack of strict change control on the project.

- All proposed changes should be assessed for impact on cost schedule and technical impact.
- A project library should be established to hold all project documentation.
- A procedure should be in place to ensure that the project team understands how to work with the project library.
- Risks and issues are the same as far as management. Action must be taken on both.
- Anyone can raise a risk or an issue, and risks and issues must always be assessed.
- Risks must be assessed for impact and likelihood.
- Issues should be assessed for impact and urgency.
- Both risks and issues should be reviewed at each project achievement meeting.

Time to Do Some Work on Monitor and Control Achievement

Answer the following questions.

1. Is there a project support office in your company or assigned to your project? If not, do you think it would be worthwhile to have one? What will you do to put one in place?

2. Do you have the disciplines of project planning, monitoring, and reporting on your project? If not, why not? Do you think it would be worthwhile to have them on your project?

3. Do you have the discipline of change control on your project? If not, why not? Do you think it would be worthwhile to have it on your project?

4. Do you have the discipline of risk and issue management on your project? If not, why not? Do you think it would be worthwhile to have it on your project?

5. Do you have the discipline of documentation management on your project? If not, why not? Do you think it would be worthwhile to have it on your project?

6. Do you use a visibility room or wall to provide information on your project? If not, why not?

CHAPTER 9

Strategy 7—Stay Focused

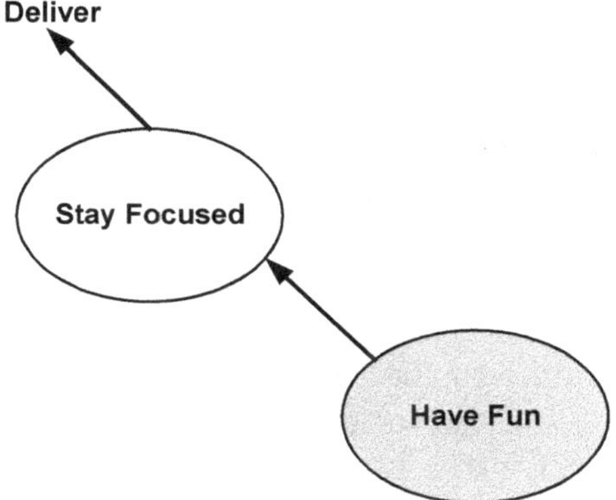

Projects Might Start Focused But...

What I have found over the years is that the people working on a project tend to lose clarity (if they had clarity) about what the project was about. They lose focus for many reasons. Some of those reasons include the following:

- Making changes the customer wants without going through change control.

- Not keeping the project documentation up to date.
- Distracted by other work they must do.
- Not sticking to the project plan.
- Disagreements with other team members.
- Distracted by non-project requirements and so on.

Ways to Help the Team Stay Focused

One strategy successful project managers practice is the strategy of staying focused. Staying focused on delivery of the agreed goals and objectives is not only important for the project manager to practice; it is also important for the entire project team and the project stakeholders. It is important, therefore, that the team stays focused on what they have committed to do and what they have committed to delivering. And there are many ways to make this happen. These include activities, such as making the project outcome visible to the team and to all stakeholders, making the individual deliverables visible to the team and to those committed to the production of those deliverables, reviewing the plan regularly, developing a motto for the project, conducting project presentations regularly for all stakeholders, and establishing a visibility room or wall.

Keeping the Goals, Objectives, and Outcome Visible

Over the years, I have used many ways to keep the project goals, objectives, and outcome visible. One way is to produce an 18 x 24-inch colorful wall poster listing the goals, objectives, and outcome. Make sure that the poster is easy to change. During the project, changes should be made to the poster to reflect agreed changes approved during the change control process if they affect the information on the poster. Put them in the elevator, on the walls, in the meeting room, and so on. Put them wherever team members and other stakeholders will notice them.

Another way is to publicize the project goals, objectives, and outcomes by printing them on a laminated project bookmark or a postcard-sized, laminated document that can be put on top of a team member's PC or goals, objectives, and outcomes mouse mat. (If your organization is still using mouse mats) The mouse mat idea is only good when we believe the goals, outcomes, and objectives will not change. It could get expensive reprinting mouse mats every time a proposed change to the outcome is approved.

Develop a project logo that typifies the goals, objectives, and outcomes. Use the logo on project documentation, letters, e-mail banners, and so on. Design and use project stationery that includes the project logo in the design. If your project is large enough, and you have or can get the budget, put the logo on T-shirts, baseball caps, coasters, and so on.

Develop a project motto that also typifies the goals, objectives, and outcomes. For instance, I was one of the program directors on the Retail Transformation Program (RTP), a major program of work for one of the large High Street banks in England. My responsibility was the Existing System Changes Program (ESCP). ESCP affected almost all the bank's existing systems.

One day, I asked the chief executive what his most important priority was in terms of information technology. He told me that his most important priority was to "keep the show on the road." In other words, keep the existing systems running while we made all RTP changes so the business could continue to operate. He said his other most important priority (I'm sure at some point, you've had more that one "most important" priority) was to deliver ESCP. We had a good think about what he said and devised the motto "keeping the road under the show." OK, people, so it's a little wet. It grew on us, and we eventually used it with our RTP logo.

Add a little fun to the project and have a contest to see who can devise the best logo and motto. Buy a bottle of champagne for the winner. Dip into your pockets if you don't have the budget on the project to do it. It is worth it. Perhaps the winner will give you a sip.

Visibility Room

Use the visibility room that we covered in the last chapter to keep the entire organization up to date on the goals, objectives and outcomes.

Project Manager Runs Interference for the Team

The picture I have in my mind when I discuss this topic is of a guard, or quarterback, or other player on an American football team. I call it American football because what people in the US call soccer, the rest of the world calls football, hence, the distinction of American football. When running some plays in American football, the responsibility of a person or persons on the team is to run in front of the person with the ball and block potential tacklers out of the way, which is called "running interference." And I see it the same way on a project. The project manager's responsibility is to run interference for the project team. By that, I mean to keep the team focused on the goals, objectives, and outcome by blocking, as much as possible, all non-project-related activities—activities that might help cause team members to lose focus.

What could some of those activities be? How about training that does not relate to the project or to the future of team members and could be put off until the project's completion or requests by others to work on anything that could affect the commitment the team member made to the project schedule or project plan. The project's completion according to the agreed plan is important here. We should focus on those activities and those activities only.

If we find that there are things we can't block from the project's way, and they start to affect the project schedule adversely, we need to tell our project board that the schedule is affected and request an approved change to the schedule, budget, and so on.

Chapter 9: Points to Remember

Points to Remember

- Projects might start clear and focused, but as they move through their life cycle, they tend to lose focus.
- Help the team stay focused by making the project's goals, objectives, and outcomes as visible as possible.
- Develop a project motto and a project logo and use them wherever you can. As an example, use them on project-specific stationery, documentation, and so on.
- Use the visibility room/wall to keep everyone focused on the goals, objectives and outcomes.
- Help the team stay focused by running interference—blocking—as much as possible, non-project-related activities.

Time to Do Some Work on Stay Focused

Answer the following questions.

1. How do you keep your project team focused on the project and what will be delivered?

____

____

____

____

____

2. Have you made the project's goals, objectives, and outcomes visible? How have you done that? If you haven't, what could you now do to make them visible?

3. Do you have a project motto and/or a project logo? If not, why not? Where would you use the logo and the motto? What do you think are the benefits of both the motto and the logo?

4. As the project manager, do you work to keep all non-project activities away from the project team? If you find it impossible to keep all non-project activities away from the team, do you keep management informed about the impact on the project schedule and budget? How do you do this? If you don't keep them informed, why not?

CHAPTER 10

Strategy 8—Have Fun

But This Is Work—What's This Having Fun Business?

Studies have shown that people learn more and are more productive when they enjoy themselves. Sure, this is work, and work is serious. What we are asked to deliver is probably also serious. But, it is the project manager's responsibility to make working on the project a positive experience for the project team members. When I say have fun, I don't mean to treat work or the project you are working on as an amusement park. What I do mean, as I always suggest to project managers, is lighten up. Don't take yourself too seriously. Team members shouldn't take themselves too seriously, either.

You need to develop the ability to make working in the highest pressure environments seem like a party and at the same time deliver to critical deadlines. You need to inspire the team through fun as well as professionalism. On many of my projects, I've appointed a Minister of Fun. The Minister of Fun must take on this role as well as his or her chief role on the project. The prime responsibility of the Minister of Fun is to devise fun things for the project team members to do.

I wish that I had devised the idea of a Minister of Fun, but I didn't. The first time I heard of a Minister of Fun being established was by Mayor William Schaefer of Baltimore, Maryland. Mayor Schaefer established the role of Minister of Fun with the overall responsibility of finding fun things for the people of Baltimore to do—and, if possible, to raise money for the city treasury while they did them.

My favorite fun idea that the Baltimore Minister of Fun devised met both the mayor's criteria. Just before Valentine's Day, the city sold potholes as valentines. You bought a pothole for, I believe, $25 or $5 for "old folks." The city filled in your pothole, drew a heart with Cupid's arrow through it with your Valentine's name in the heart, took a picture of it, put it in a Valentine's card, included the address of your pothole, and sent it to you to give to your Valentine. Great fun idea, and the city sold enough potholes to repair their streets.

There was also Think Pink Day, but I won't go in to that. You need to read about William Schaefer and his exploits yourself.

But the idea for a Minister of Fun is a good one, and as I've said, I have implemented it on many of my projects and it accomplished what I wanted it to do. Perhaps the Minister of Fun should have a slot at the weekly achievement meeting. Help him or her out by brainstorming fun ideas to do for the project.

Fun things will help lighten up the project team's spirits to do more, faster.

I've included some ideas below that you might be able to use or might help stimulate your thinking to find other ways to have fun at work. I suggest you get the book *301 Ways to Have Fun at Work* by Dave Hemsath. There are some great ideas in the book.

Sponsor a show-and-tell day where project team members bring to the achievement meeting something they are interested in and would like to share with the team. Sponsor a contest to name the project. The best name chosen by team vote wins a bottle of champagne.

Have a baby picture day with all the team bringing pictures of themselves as babies and have a contest to see who can match the most baby pictures with the project team member.

Most people on our project today seem to juggle loads of things at work. Well, surprise your team by bringing a professional clown/juggler to teach them how to juggle. I did this once at a staid financial institution in the City of London. When one of my teams complained that they were having to juggle loads of different activities to keep things going, I hired a clown from Covent Garden to come in, and instead of having our weekly achievement meeting, this clown taught our project team how to juggle. It worked. It loosened up the atmosphere on the project not only during the lesson, but the new atmosphere also lasted for the rest of the project.

Chapter 10: Points to Remember

Points to Remember

- Loosen up the atmosphere on your project by providing a positive experience for people working on your project.
- Appoint a Minister of Fun on your project with the prime responsibility of finding fun things to do. Do it. It works.

Time to Do Some Work on Have Fun

Answer the following questions.

1. Are your projects fun to work on? If not, what could you do to make them more enjoyable for the project team?

____

____

____

____

____

____

2. Does your project provide a positive experience for members of the project team?

____

____

____

____

____

3. Do people work on your project because they want to or because they have to? Which would you prefer?

___

___

___

4. Will you appoint a Minister of Fun? If not, why not?

___

___

___

___

___

___

___

5. Do you think of fun things for the project team to do?

___

Have you enjoyed this book? If you have or if you haven't, please let me know at richard@richardmorreale.com. After I published my first book, Over 50 Killer Ideas for Delivering Successful Systems, I received loads of emails commenting on the book. One day I received two comments. The first one said what a great book it was and how much they enjoyed it and the second said what a lousy book it was. I took a page out of one of Dr. Wayne Dyers programs and forwarded each to the other with the comment, "You might be right".

Thank you very much and I look forward to hearing from you.

About the Author

Richard is the founder and CEO of Inspiration Systems Inc. He has managed projects, conducted training, led workshops, rescued projects that were in desperate trouble and spoken at conferences, company meetings and associations in the US, the Middle East, Africa, Asia, the UK and other European countries.

His project and program experience ranges from working as part of the Apollo Program Team, helping to put people on the moon (and getting them back) to working as part of the team that computerized the UK Income Tax System. He also led the rescue and delivery of a $450M Program for the 43 Police Forces in England and Wales and directed programs of work for some of the largest companies in the world!

He has international experience of founding and leading the growth of a successful Project Management Consultancy in the UK, managing organizations of up to 1200 people for various clients and working in a number of different organizational sectors including Telecommunications, UK Local and Central Government, Financial Services, the US Government, Utilities, Transportation, the IT Industry and Retail.

He is a member of the National Speakers Association (NSA) and on the Board of the NSA/ Carolinas Chapter.

To book Richard or to make welcomed comments to *Top Gun Project Managers,* contact him at Richard@RichardMorreale.com or on 336-322-1320.

Richard's mission is to provide people with information that they can use to make a better life for themselves, their families and their companies.

Did you like this book?

If you enjoyed this book, you will find more interesting books at

www.MMPubs.com

Please take the time to let us know how you liked this book. Even short reviews of 2-3 sentences can be helpful and may be used in our marketing materials. If you take the time to post a review for this book on Amazon.com, let us know when the review is posted and you will receive a free audiobook or ebook from our catalog. Simply email the link to the review once it is live on Amazon.com, with your name, and your mailing address—send the email to orders@mmpubs.com with the subject line "Book Review Posted on Amazon."

If you have questions about this book, our customer loyalty program, or our review rewards program, please contact us at info@mmpubs.com.

Over 50 Killer Ideas for Delivering Successful Projects

by Richard Morreale

Over the past 30 years, companies have spent millions (if not billions) of dollars on Project Management tools, techniques, methods, processes, procedures, and training to help their managers successfully plan, organize, monitor, and control their projects. After all of this investment, however, the same percentage of projects are failing today as there were 30 years ago.

In this book, Richard Morreale presents over 50 killer ideas that will help your projects become successful: delivering on time, within budget, and meeting your client's expectations. This energetic and dynamic speaker brings life to this subject, sharing the lessons he's learned in his decades of experience managing high-profile projects for government and industry.

ISBN: 9781897326756 (paperback)

Also available in ebook and audiobook formats. Order from your local bookseller, Amazon.com, or directly from the publisher at **http://www.mmpubs.com**

Thinking on Purpose for Project Managers: Outsmarting Evolution

When you're facing down a lion on the open savannah, automatic reactions hardwired into your system through eons of evolution can save your life. However, when you're trying to impress the CEO across a boardroom table, those same responses can cost you big time.

So, how do you overcome your automatic reactions, retrain your brain, and outsmart evolution? By learning and using the techniques revealed in *Thinking on Purpose for Project Managers: Outsmarting Evolution.*

ISBN: 9781554890255 (paperback)

Also available in ebook formats. Order from your local bookseller, Amazon.com, or directly from the publisher at **http://www.mmpubs.com**

Project Lessons from The Great Escape (Stalag Luft III)

While you might think your project plan is perfect, would you bet your life on it?

In World War II, a group of 220 captured airmen did just that – they staked the lives of everyone in the camp on the success of a project to secretly build a series of tunnels out of a prison camp their captors thought was escape proof.

The prisoners formally structured their work as a project, using the project organization techniques of the day. This book analyzes their efforts using modern project management methods and the nine knowledge areas of the *Guide to the Project Management Body of Knowledge* (PMBoK).

Learn from the successes and mistakes of a project where people really put their lives on the line.

ISBN: 9781895186802 (paperback)

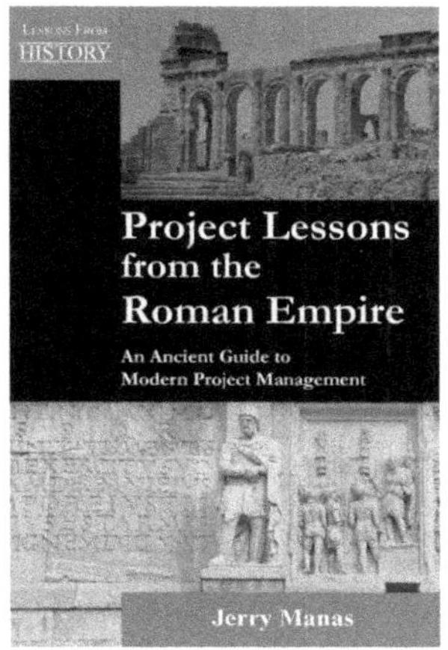

Project Lessons from the Roman Empire: An Ancient Look at Modern Project Management

The leaders of the Roman Empire established many of the organizational governance practices that we follow today, in addition to remarkable feats of engineering using primitive tools that produced roads and bridges which are still being used today, complex irrigation systems, and even "flush toilets." Yet, the leaders were challenged with political intrigue, rebelling team members, and pressure from the competition. How could they achieve such long-lasting greatness in the face of these challenges?

In this new addition to the Lessons from History series, join author Jerry Manas as he takes you on a journey through history to learn about project management the Roman way. Discover the 23 key lessons that can be learned from the successes and failures of the Roman leadership, with specific advice on how they can be applied to today's projects.

Looking at today's hottest topics, from the importance of strategic alignment for your projects through to managing transformational change and fostering work/life balance while still maintaining overall performance, you'll find that the Romans already faced-and conquered-these challenges two thousand years ago. Read this intriguing book to learn how they did it.

ISBN: 9781554890545 (paperback)

The History of Project Management

The Pyramid of Giza, the Colosseum, and the Transcontinental Railroad are all great historical projects from the past 4,000 years. When we look back, we tend to look at these as great architectural or engineering works. Project management tends to be overlooked, and yet its core principles were used extensively in these projects.

This book takes a hard look at the history of project management and how it evolved over the past 4,500 years. It shows that "modern" project management practices did not just appear in the past 100 years but have been used — often with a lot of sophistication — for thousands of years.

Examining archaeological evidence, artwork, and surviving manuscripts, this book provides evidence of how each of the nine knowledge areas of project management (as shown in PMI's PMBoK® Guide) have been practiced throughout the ages. The book covers the period from the construction of the ancient pyramids up to the 1940s. A future companion volume will cover more recent developments during the war years and the technology boom leading up to the present.

As readers explore the many case studies in this book, they will discover fascinating details of innovative projects that produced our most famous landmarks and voyages of discovery.

ISBN: 9781554890965 (hardcover)

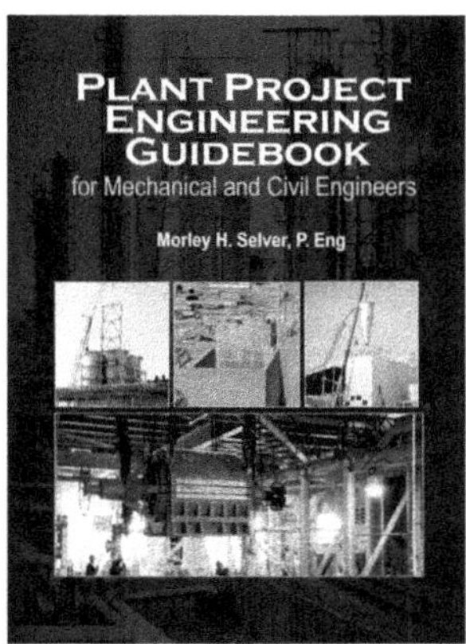

Plant Project Engineering Guidebook for Mechanical and Civil Engineers

This book is for plant project engineers who want an insight into the main facets of project engineering in North America. Engineers are guided through work procedures and methods for plant projects in the oil and gas, pulp and paper, board and recycling industries and that can be applied to other industries.

The book contains 168 practical guidelines, 18 case histories and 30 sample forms to describe the project authorization process including budgets and estimating, information on how to control the office and drawing functions, a description of the bidding and procurement process, the basics of construction management, contracts, commissioning procedures, startup and training.

This is the stuff you wish they would have taught at university, but didn't.

ISBN: 9781554890996 (hardcover)

Also available in ebook formats. Order from your local bookseller, Amazon.com, or directly from the publisher at **http://www.mmpubs.com/**